T0310079

CHEMISTRY AND PHYSICS OF MECHANICAL HARDNESS

CHEMISTRY AND PHYSICS OF MECHANICAL HARDNESS

John J. Gilman

WILEY

A JOHN WILEY & SONS, INC., PUBLICATION

Library of Congress Cataloging-in-Publication Data:

Gilman, John J. (John Joseph)
   Chemistry and physics of mechanical hardness / John J. Gilman.
      p. cm. – (Wiley series on processing of engineering materials)
   "A Wiley-Interscience publication."
   Includes bibliographical references and index.
   ISBN 978-0-470-22652-0 (cloth)
  1. Hardness. 2. Strength of materials. I. Title.
   TA418.42.G55 2009
   620.1′126–dc22

                                    2008038594

Printed in the United States of America.
10 9 8 7 6 5 4 3 2 1

TABLE OF CONTENTS

PREFACE

For the structural applications of materials, there is no more useful measurable property than mechanical hardness. It quickly and conveniently probes the strengths of materials at various scales of aggregation. Firstly, it does this at the human scale (Brinell hardness—millimeters to centimeters). Secondly, it does so at a microscopic scale (Vickers microhardness—1 to 100 microns). And thirdly, it does so at a "nanoscale" (nanoindentation—10 to 1000 nanometers).

For millenia, hardness has been used to characterize materials; for example, to describe various kinds of wood ranging from soft balsa wood to hard maple and ironwood. Mineralogists have used it to characterize differing rocks, and gemologists for the description of gems. Ceramists and metallurgists depend on it for classifying their multitude of products.

Hardness does not produce a complete characterization of the strengths of materials, but it does sort them in a general way, so it is very useful for "quality control"; for the development of new materials; and for developing prototypes of devices and processes. Furthermore, mechanical hardness is closely related to chemical hardness, which is a measure of chemical bond stability (reactivity). In the case of metals the connection is somewhat indirect, but nevertheless exists.

The principal intention of the present book is to connect mechanical hardness numbers with the physics of chemical bonds in simple, but definite (quantitative) ways. This has not been done very effectively in the past because the atomic processes involved had not been fully identified. In some cases, where the atomic structures are complex, this is still true, but the author believes that the simpler prototype cases are now understood. However, the mechanisms change from one type of chemical bonding to another. Therefore, metals, covalent crystals, ionic crystals, and molecular crystals must be considered separately. There is no universal chemical mechanism that determines mechanical hardness.

There have been a number of past attempts to unify hardness measurements but they have not succeeded. In several cases, hardness numbers have been compared with scalar properties; that is, with cohesive energies (Plendl and Gielisse, 1962) or bulk moduli (Cohen, 1988). But hardness is not based on scalar behavior, since it involves a change of shape and is anisotropic. Shape changes (shears) are vector quantities requiring a shear plane, and a shear direction for their definition. In this book, the fact that plastic deformation is a shear process mediated by the motion of dislocations is the basis of the discussion.

In order to treat hardness quantitatively, it is essential to identify the entities (energies) that resist dislocation motion as well as the virtual forces (work) that drive the motion. These are the "ying and yang" of hardness. They are very different in pure metals as compared with pure covalent solids, and still different in salts and molecular crystals.

The author has struggled to develop an understanding of plastic deformation (and therefore hardness) for several decades. This has not been a straightforward task because the literature of the subject has been, and still is, confused in part (i.e., wrong). Only gradually has the author come to realize that textbooks are not necessarily correct in their interpretations of phenomena. Even experts sometimes accept misinterpretations of phenomena that, through repetition, have become gospel.

The subject of plastic deformation has suffered from attempts to interpret macroscopic behavior without adequate microscopic (and nanoscopic) information. This will always be the case to some extent, but it needs to be minimized. Also, since the size scale of dislocations is atomic, Heisenberg's principle and its implications must be considered in order to understand plastic deformation and, therefore, hardness.

REFERENCES

M. L. Cohen, "Theory of Bulk Moduli of Hard Solids," Mater. Sci. & Eng. A, **105–106**, 11 (1988).

J. N. Plendl and P. J. Gielisse, "Hardness of Nonmetallic Solids on an Atomic Basis," Phys. Rev., **125**, 828 (1962).

1 Introduction

1.1 WHY HARDNESS MATTERS (A SHORT HISTORY)

A most characteristic property of a solid is its hardness. This ranges from very soft (talc) to very hard (diamond). Although hardness is an important general characteristic of materials, it also has great utility. It determines the resistances of surfaces to wear. It determines the effectiveness of all sorts of tools used for cutting everything from textiles to hard rocks. A closely related function is the polishing of gems, mirrors, lenses, and the like. It is an index of the strengths of materials; particularly metals. Geologists find it useful for identifying minerals; and it plays a key role in geophysical phenomena such as meteor impacts. Unfortunately it can also be very destructive in military ordnance. A property of more ubiquitous importance is hard to find.

The range of hardness numbers, measured in kilograms per square millimeter, is large. It runs from one for a soft material like KI to about ten thousand for the hardest material—diamond. In other words, it has a range of about four orders of magnitude.

As a result of its utility, mechanical hardness has been highly prized for millennia. It has often played a key role in the progression of civilization because it has enabled progressively more sophisticated devices and machines to be constructed. Initially the hardest available materials were rock, bone, and wood. Bone tools from 19,000 BCE have been found. The search for improved hardness extends back to char hardened wooden-spears as old as 120,000 BCE (Bunch and Hellemans, 2004). The first advanced material was probably flint which fractures conchoidally, so controlled fracture can give it atomically sharp edges (Wikipedia, 2006). Flint was a considerable improvement on obsidian (volcanic glass), starting in the Middle Paleolithic Age (≈300,000 to 30,000 years ago). Imagine the improvement that flint arrowheads made, compared with hard wood. Hardness was not the only factor that determined technological progress, but was a key factor. Technology often waited for improved hardness to become available before new technology could be introduced.

For example, consider copper. Native copper could be worked into various shapes, but was too soft for making tools and swords. Copper smelted from ore was also relatively soft. This changed when it was discovered that calamine ore (mixed carbonates and silicates of zinc), when smelted together with copper ore makes brass (a Cu-Zn alloy much harder than copper alone). Further improvement came when tin became available, and could be mixed with copper to make bronze. Syrians did this in about 3000 BCE, thereby ending the Stone Age and beginning the Bronze Age.

One of the most important inventions of the late Stone Age was the wheel. It reduced the force needed to move objects by a factor of at least 100. But the first wheels were made of wood. An enormous advance came when wheels could be rimmed with bronze.

Since tin is much less volatile than zinc, bronzes are more versatile than brasses. The low volatility of tin allowed bronze alloys to be cast and forged into various shapes, including large bells. These formed the basis of the first large scale communication systems. Every settlement had its own bell tower for communicating with its residents, and with other settlements.

Bronzes with a variety of hardnesses and other properties can be made depending on the tin content, and by adding other elements as well as tin. These have included arsenic, magnesium, calcium, phosphorous, and antimony. The resulting bronzes had (have) a large variety of applications. Because it forms a protective oxide, bronze is excellent for fittings on boats where it resists seawater corrosion. Its friction coefficient is small so it makes excellent bearings and gears. Alloyed with phosphous, it is useful for springs. Since it fills casting molds precisely, it is used extensively for statues and other art objects.

In parallel with the development of the art of smelting metallic ores, the art of heating various oxides and silicates to make artificial rocks (ceramics) was being slowly developed. The first evidence of this development has been found in Moravia, dating from 28,000 BCE. Clay pottery was being manufactured in Egypt by about 5000 BCE, and glass glazes were applied to their surfaces from about 12,000 BCE. Glass shapes date from 7,000 BCE in Egypt. The availability of clay pottery revolutionized the transport of water, wine, and other liquids.

Special glass compositions have been invented, using a variety of oxides together with silica to obtain special characteristics. An early one was lead glass in England in 1624. For improving glass lenses, Zeiss, Inc. invented high boron glass in 1884, and Abbe-Schott developed high barium glass in 1888. A special boron glass (Pyrex) was developed at the Corning Glass Works in 1915. Precipitation-hardened glasses were developed at the same place during the mid-twentieth century. High purity, high strength glass fibers for optical communication systems began to be made in the 1970–80 period.

A dramatic change in the hardnesses of metallic materials occurred when the smelting of iron was invented in Egypt (4000 BCE). By 1550 BCE, it could be forged into wrought iron which considerably improved its properties.

By adding relatively large amounts of carbon (several weight per cent) to iron, it was found in China (500 BCE) that large and complex cast-iron shapes could be made readily. Through chill-casting, cast-iron becomes very hard (albeit brittle); this gives it considerable wear resistance.

Although pure iron is less hard than some bronzes, it was considerably hardened by converting it to pearlitic steel (iron + a small amount of carbon) (India, 500BCE). By adding more carbon (up to 1 wt.%), and other selected metals, plus heating, quenching, and tempering (reheating), iron alloys yielded very hard, tough martensitic steels.

Without these advances in hard, strong materials; based on abundant, and therefore low-cost iron ore, there could have been no industrial revolution in the nineteenth century. Long bridges, sky-scraper buildings, steamships, railways, and more, needed pearlitic steel (low carbon) for their construction. Efficient steam engines, internal combustion engines, turbines, locomotives, various kinds of machine tools, and the like, became effective only when key components of them could be constructed of martensitic steels (medium carbon).

The civilian advances were accompanied, and often led, by advances in military ordnance. Iron and steel became the basis of swords, spears, arrows, guns, cannon, armor, tanks, warships, and more. In fact, the motivation for inventing and developing new hard materials was often the desire for improved military ordnance. This continues with searches for better body armor, and the inverse searches for more penetrating projectiles.

An important sub-division of the industrial revolution was the discovery by Moissan in 1906 that carbon forms exceedingly hard tungsten carbide (WC) crystals. In 1928, workers at Krupp, Germany found that WC crystals can be cemented with cobalt metal to make aggregates that were unparalleled tools for cutting steel (Riedel, 2000, p. 481). Other hard compounds, such as silicon carbide (SiC) and aluminum oxide (Al_2O_3) are also used for cutting other materials. The hardest crystals of all, diamond and cubic boron nitride (BN), are very useful for cutting rock as well as steel, in the case of BN. Diamond is not useful for cutting steel because, being carbon, it reacts with iron. It began being used as a tool as early as 300 BEC (Riedel, 2000).

With the advent of aeronautics, aluminum alloys allowed major advances such as the monocoupe design. The first all metal airplane was the Junkers J-1 (1917). Pure aluminum is light in weight, but too soft for constructing aircraft, so it is hardened by adding to it copper, magnesium, and other metals. During heat treatments, these form precipitate particles that harden the alloy (e.g., particles of the compound, $CuAl_2$). This process is called age-hardening. It is the approach used for the alloy known as Duralumin which was invented by Alfred Wilm in Germany in about 1909, and has been a standard construction alloy for many years.

Aeronautics also stimulated the development of superalloys, largely based on nickel. They hold their strength (hardness) at very high temperatures. Efficient aircraft-turbine engines could not be constructed without them. These engines have allowed mass commercial air transportation to develop.

The weakest parts of superalloys are the grain boundaries between the crystals. A desire to eliminate these boundaries led F. VerSnyder to fabricate turbine blades from very large individual crystals of nickel–aluminum alloys. This material consists mostly of the compound, Ni_3Al, and because of its lack of grain boundaries, the monocrystalline form has remarkable creep resistance and high temperature fracture resistance. The durability of these monocrystalline turbine blades has substantially reduced the costs of operating large jet airplanes by extending the time between repair operations. They have also reduced fuel consumption by increasing the maximum allowable operating temperatures.

In addition to mechanical devices, optical devices have benefited from improvements in the hardnesses of materials. A familiar example is scratch-proof lenses for eye glasses. Scratch-resistant aircraft windscreens are also important. Less familiar, but more impressive, is the importance of hardness in solid-state lasers. Early in the history of solid-state lasers, it was discovered that the best host material for the active fluorescent atoms is a very hard garnet (yttrium aluminum garnet—YAG). The standard fluorescent centers are neodynium atoms added to the garnet. YAG garnet is exceptionally rugged and hard at very high temperatures. Because of the intensity (energy/sec.cm^2) of the light beam in the lasing material, large electric fields tend to decompose it. Therefore, both its optical and mechanical properties contribute to desirable performance.

High pressure scientific research is an area of science that has benefited from high hardness. Here, individual diamonds are used as pressure vessels to contain specimens at ultra-high pressures (millions of times atmospheric pressure).

Because of the simplicity of doing scratch tests, hardness has been an important diagnostic tool for mineralogists and prospectors by helping them to identify various rocks and minerals.

Since it measures the susceptibility of materials to plastic deformation (as contrasted with elastic deformation), hardness is very important for diagnosing the mechanical state of a material, in particular toughness. Purely elastic materials are brittle. Plasticity, by blunting cracks and other defects, allows metals and, to some extent ceramics, to tolerate small flaws and thereby become malleable and tough.

An illustration of the impact that improved hardness has had on technology is presented in Figure 1.1. This shows the dramatic increase in performance of machine cutting tools (lathes, milling machines, saws, drills, and the like) as the tools became harder. It also shows how very fast cutting speeds have become. The top cutting speed (≈ 5000 m/sec) is about 16% of the speed of sound in air!

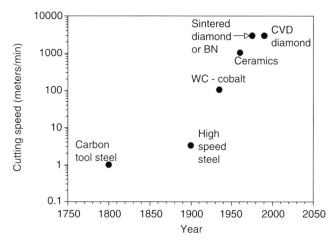

Figure 1.1   Improvement in cutting speeds with tool hardness. During two centuries of development machine tool performance increased by a factor of about 5000. Adapted from Riedel, 2000, p. 550.

1.2   PURPOSE OF THIS BOOK

Crystals of high purity metals are very soft, while high purity diamond crystals are very hard. Why are they different? What features of the atomic (molecular) structures of materials determine how hard any particular crystal, or aggregate of crystals, is? Not only are crystals of the chemical elements to be considered, but also compounds and alloys. Glasses can also be quite hard. Is it for similar reasons? What about polymeric materials?

Many decades ago, with the advent of convincing atomic theory, it was thought that a universal model for hardness could be found. This is not the case given the present state of solid-state physics. Much of physics, and therefore chemistry, is based on interactions between pairs of particles. This is adequate for understanding changes of sizes of objects, but hardness involves changes of shape, and this requires more complex interactions.

However, It has been found that in many cases, simple models of the properties of atomic aggregates (monocrystals, polycrystals, and glasses) can account quantitatively for hardnesses. These models need not contain disposable parameters, but they must be tailored to take into account particular types of chemical bonding. That is, metals differ from covalent crystals which differ from ionic crystals which differ from molecular crystals, including polymers. Elaborate numerical computations are not necessary.

The presentation here attempts to provide—for materials scientists, metallurgists, ceramists, chemists, and physicists—knowledge of how hardness is related to other properties, and to the building blocks of everyday matter— atoms and electrons; that is, what information is contained in hardness measurements. The emphasis is on physical concepts so the general picture may be grasped and appreciated by most readers. Various materials types are discussed in individual chapters. Some chapters on general principles integrate the whole.

This is, by no means, the first attempt to relate hardness to other more precisely defined properties. As mentioned in the Preface, Plendl and Gielisse (1962) studied correlations between hardness and cohesive energies per atomic volume. Both quantities have the same units (energy/volume). These correlations are successful, but not completely. The shortcoming is that cohesive energy is a measure of the energy needed to *separate* atoms, but hardness is not a measure of this. Hardness is a measure of the energy needed to *shear* pairs of atoms; that is, to break chemical bonds by shearing them.

Other authors have studied other correlations. Two are Povarennykh (1964), and Goble and Scott (1985). The latter emphasized compressibility (inverse bulk modulus) as did Beckmann (1971). The bulk modulus is not a reliable measure for the same reason as the cohesive energy. It is volume dependent rather than shear dependent. Still another attempt to correlate hardness and compressibility was that of Yang et al. (1987). This was followed by a proposal by Liu and Cohen (1990) that hardness and bulk moduli are related. This proposal was refuted by Teter (1998) who showed that hardness values correlate better with shear moduli than with bulk moduli.

A measure of shear strength is the shear modulus. For covalent crystals this correlates quite well with hardness (Gilman, 1973). It also correlates with the hardnesses of metals (Pugh, 1954), as well as with ionic crystals (Chin, 1975). Chin has pointed out that the proportionality number (VHN/C_{44}) depends on the bonding type. This parameter has become known as the Chin-Gilman parameter.

The variation of the Chin-Gilman parameter with bonding type means that the mechanism underlying hardness numbers varies. As a result, this author has found that it is necessary to consider the work done by an applied shear stress during the shearing of a bond. This depends on the crystal structure, the direction of shear, and the chemical bond type. At constant crystal structure, it depends on the atomic (molecular volume). In the case of glasses, it depends on the average size of the disorder mesh.

There are at least four types of chemical bonding. Some crystals have open atomic structures, while others are close-packed. Also, many crystals are anisotropic. Therefore, although making hardness measurements is relatively simple, understanding the measured values is not simple at all.

Attempts to understand hardness from first principles have resulted in empirical equations that represent good curve fitting, but yield relatively little understanding (Gao, 2006).

1.3   THE NATURE OF HARDNESS

Hardness is a measure of the ease with which solids can be plastically deformed. This depends on the mobilities of dislocations, their multiplication, and their interactions. Dislocation speeds vary from Angstroms per second to 10^{13} Å/sec. Their concentrations vary from zero to about 10^{12} lines/cm^2 and the interaction possibilities number at least the squares of their concentrations. Fortunately, there are some limiting cases in which a few factors dominate the behavior.

The mobilities of dislocations are determined by interactions between the atoms (molecules) within the cores of the dislocations. In pure simple metals, the interactions between groups of adjacent atoms depend very weakly on the configuration of the group, since the cohesive forces depend almost entirely on the local electron density, and are of long range.

In covalently bonded crystals, the forces needed to shear atoms are localized and are large compared with metals. Therefore, dislocation motion is intrinsically constrained in them.

Ionically bonded crystals contain both long-range and short-range bonding forces because like ions repel each other, while unlike ones attract.

Thus, in simple metals, interactions between dislocations rather than interactions between atoms, are most important. The hardnesses of metals depend on deformation hardening (dislocation interactions) rather than individual mobilities. The elastic resistance to shear plays a dominant role because it is directly involved with dislocation mobility.

Since hardness and the shear modulus are usually proportional, the factors that determine the shear moduli need to be understood. The shear moduli are functions of the local polarizability and this depends on the valence electron density, as well as the energy needed to promote a valence electron to its first excited state. The latter depends on the strength of the chemical bond between two atoms. This will be discussed in more detail in Chapter 3.

Hardness is a somewhat ambiguous property. A dictionary definition is that it is: "a property of something that is not easily penetrated, spread, or scratched." These behaviors involve very different physical mechanisms. The first relates to elastic stiffness, the second to plastic deformation, and the third to fracturing. But, for many substances, the mechanisms of these are closely related because they all involve the strength of chemical bonding (cohesion). Thus discussion of the mechanism for one case may provide some understanding of all three.

The four rather distinct forms of chemical bonding between atoms are: metallic, ionic, covalent, and dispersive (Van der Waals). All of them are subtopics of quantum electrodynamics. That is, they are all mediated by electronic and electromagnetic forces. There are also mixed cases, as in carbides and other compounds, where both metallic and covalent bonding occur.

The principal type of hardness to be discussed here is indentation hardness in which a diamond of a standard shape is impressed into a specimen surface.

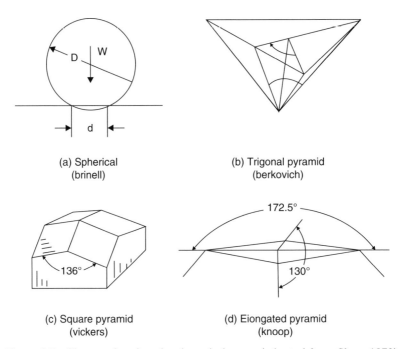

(a) Spherical (brinell)	(b) Trigonal pyramid (berkovich)
(c) Square pyramid (vickers)	(d) Eiongated pyramid (knoop)

Figure 1.2   Shapes of various hardness indenters (adapted from Shaw, 1973).

The shape is usually either: a sphere (Brinell, and Rockwell B or C); a square pyramid with apex angle = 135° (Vickers); a trigonal pyramid (Berkovich); or an elongated four-sided pyramid (Knoop). (See Figure 1.2). For quality control in manufacturing operations, semi-automatic Rockwell machines, and their various indenters, are also useful.

A fixed force is applied to the axis of the indenter which makes an irreversible indentation into the specimen's surface. The projected length, or area, of this indentation is measured, and the ratio of the applied load to this projection is formed to obtain the hardness number which has the dimensions of stress (also expressable as energy/volume). The sizes of the indentations vary, depending on the indenter's shape and the amount of load applied to it. The size range is from macro- (millimeters), through micro- (microns), to nano- (nanometers).

There are other, less commonly used, methods for measuring hardness. One is an impact method in which an indenter is dropped from a known height onto a specimen, and either the size of the indentation, or the coefficient of restitution, is measured. Another is the pendulum method in which a rocking pendulum is applied to a specimen surface. The damping of the pendulum's oscillations is a measure of the hardness. Still another is Moh's scratch method in which the ability of one specimen to scratch another is observed. These methods are described in various books (McColm, 1990), but only the

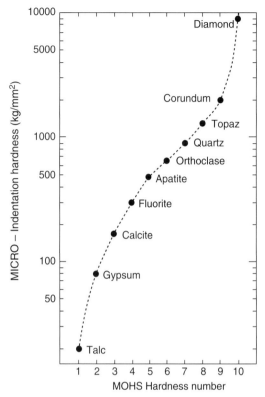

Figure 1.3 Correlation between the Moh scratch hardness and Vickers indentation hardness scales.

Vickers indentation method will be pursued here (in both its micro- and nano-manifestations).

However, to develop some intuitive sense of hardness it is useful to consider the Moh scratch hardness scale. This is a rank-file scale consisting of ten levels. Each level has been assigned to a particular mineral such that the mineral at level n is capable of scratching the one at level $(n-1)$. The mineral at the lowest level (designated 1) is talc, and the highest (designated 10) is diamond.

Figure 1.3 compares the Moh scratch scale with the more quantitative Vickers scale (Gilman, 1973). Clearly the two scales are not linearly related. Each has its own realm of application. For brittle minerals, and similar materials, the Moh scale is most useful. For ductile materials like metals the Vickers indentation scale can detect small differences more readily. Note that the range on the Vickers scale is large; about 1000, while range of the Mohs scale is about 10.

REFERENCES

G. Beckmann, "Ueber den Zusammenhang zwischen Kompressibilitat und Harte Von Mineralen und Nichtmetallischen Kristallinen Substanzen," Kristall und Technik, **6**, 109 (1971).

B. Bunch and A. Hellemans, *A History of Science and Technology*, Houghton Mifflin Co., Boston, USA (2004).

G. Y. Chin, "Strong and Hard Solids," Trans. Amer. Crystallographic Assoc., **11**, 1 (1975).

F. Gao, "Theotetical Model of Intrinsic Hardness," Phys. Rev. B, **73**, 132104 (2006).

J. J. Gilman, "Hardness—A Strength Microprobe," Chapter 4 in *The Science of Hardness Testing and Its Research Applications*, p.65, Edited by J. H. Westbrook and H. Conrad, American Society for Metals, Metals Park, Ohio, USA (1973).

R. J. Goble and S. D. Scott, "The Relationship between Mineral Hadness and Compressibility (0r Bulk Modulus)," Canadian Mineralogist, **23**, 273 (1985).

A. Y. Liu and M. L. Cohen, "Structural Properties and Electronic Structure of Low-Compressibility Materials: β-silicon Nitride and Hypothetical Carbon Nitride (β-C3N4)," Phys. Rev. B, **41**(15), 10727–34 (1990).

I. J. McColm, *Ceramic Hardness*, Plenum Press, New York, NY, USA (1990).

J. N. Plendl and P. J. Gielisse, "Hardness of Nonmetallic Solids on an Atomic Basis." Phys. Rev., **125**, 828 (1962).

A. S. Povarennykh, Chaps. 26, 27, 28, and 29 in *Aspects of Theoretical Mineralogy in the USSR*, Trans. by M. H. Battey and S. I. Tomkeieff, The Macmillan Company, New York, USA (1964).

S. F. Pugh, "Relations between the Elastic Moduli and the Plastic Properties of Polycrystalline Metals," Phil. Mag., **43**, 823 (1954).

R. Riedel (Editor), *Handbook of Ceramic Hard Materials—Vols. 1 and 2*, WILEY-VCH Verlag, Weinheim, Germany (2000).

M. C. Shaw, "The Fundamental Basis of the Hardness Test," Chapter 1 in *The Science of Hardness Testing and Its Research Applications*, p. 1, Edited by J. H. Westbrook and H. Conrad, American Society for Metals, Metals Park, Ohio, USA (1973).

D. M. Teter, "Computational Alchemy: The Search for New Superhard Materials," MRS Bulletin, **23**, 22 (1998).

Wikipedia, entry under Paleolithic (2006).

W. Yang, R. G. Parr, and L. Uytterhoeven, "New Relation between Hardness and Compressibility of Minerals," Phys. & Chem. Minerals, **15** (2), 191 (1987).

2 Indentation

2.1 INTRODUCTION

The deformation of a specimen during indentation consists of two parts, elastic strain and plastic deformation, the former being temporary and the latter permanent. The elastic part is approximately the same as the strain produced by pressing a solid sphere against the surface of the specimen. This is described in detail by the Hertz theory of elastic contact (Timoshenko and Goodier, 1970).

At the instant of contact between a sphere and a flat specimen there is no strain in the specimen, but the sphere then becomes flattened by the surface tractions which creates forces of reaction which produce strain in the specimen as well as the sphere. The strain consists of both hydrostatic compression and shear. The maximum shear strain is at a point along the axis of contact, lying a distance equal to about half of the radius of the area of contact (both solids having the same elastic properties with Poisson's ratio = 1/3). When this maximum shear strain reaches a critical value, plastic flow begins, or twinning occurs, or a phase transformation begins. Note that the critical value may be very small (e.g., in pure simple metals it is zero); or it may be quite large (e.g., in diamond).

The inelastic response (flow, twinning, or phase change) continues under a given applied force on the sphere until the increasing indentation area causes the maximum shear stress to drop below a critical value which is typically determined by the amount of deformation-hardening that occurs during the plastic indentation. The idea that is still found in the literature of the subject (including text books) is that the plastic indentation process can be described adequately by means of the continuum theory of plasticity, that is, in terms of slip-line fields and "yield" stresses. This leads to the idea of the hardness number, $H \approx 3Y$ where Y is the yield stress of a material (Tabor, 1951). Although this equation has been empirically shown to be approximately true for metals including steel (Figure 2.1), it is wrong by an order of magnitude for non-metals (Westbrook, 1958) where the numerical coefficient is approximately 35 instead of 3.

In addition to the discrepancy between observed and postulated ratios of hardness numbers and yield stresses, the observed plastic deformation

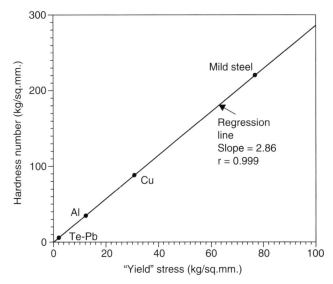

Figure 2.1   Data of Tabor (1951) comparing hardness numbers and "yield" stresses of some typical metals. The correlation is excellent and the slope is approximately three.

patterns under indenters do not agree (in general) with the slip-line fields given by continuum theory as discussed in some detail by Hill (1950). The discrepancy was first emphasized by Shaw (1973) and has been carefully studied by Chaudhri (2004).

The plastic deformation patterns can be revealed by etch-pit and/or X-ray scattering studies of indentations in crystals. These show that the deformation around indentations (in crystals) consists of heterogeneous rosettes which are qualitatively different from the homogeneous deformation fields expected from the deformation of a continuum (Chaudhri, 2004). This is, of course, because plastic deformation itself is: (a) an atomically heterogeneous process mediated by the motion of dislocations and (b) mesoscopically heterogeneous because dislocation motion occurs in bands of plastic shear (Figure 2.2). In other words, plastic deformation is discontinuous at not one, but two, levels of the states of aggregation in solids. It is by no means continuous. And, it is by no means time independent; it is a flow process.

Plastic deformation is mediated at the atomic level by the motion of dislocations. These are not particles. They are lines. As they move, they lengthen (i.e., they are not conserved). Therefore their total length increases exponentially. This leads to heterogeneous shear bands and shear instability.

In retrospect, it should not be surprising that a time independent theory modeled after elasticity theory does not apply to a plastic flow process. Elastic deformation is conservative with the work done on the material stored as elastic strain energy. Plastic deformation is non-conservative with the work done on the material dissipated as heat, or converted into internal defects

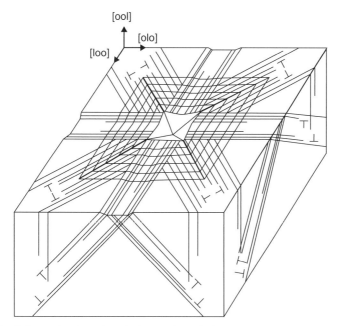

Figure 2.2   Schematic shear bands (rosette) and diagonal cracks at an indentation in MgO (after Armstrong and Wu, 1978).

(dislocations, vacancies, interstitials, changes in internal ordering, etc.) Plastic flow is more akin to viscous fluid flow than to elastic deformation. It is, in fact, quite time dependent. This, plus the fact that it is non-conservative, means that the constitutive equations that describe it must be kept separate from those that describe elastic deformations. Apples should not be mixed with oranges!

In this book, elastic strain and plastic deformation will be differentiated by both words and symbols. Elastic strain is given the usual symbols: ε and γ for extensional and shear elastic strains, respectively. For plastic shear deformation. δ will be used. ε and δ are physically different entities. ε and γ are conservative quantities which store internal energy. δ is not conservative. The work done to create it is dissipated as heat and structural defects.

The three inelastic processes (flow, twinning, and phase changes) all require the shearing of atomic neighbors, so they all tend to occur at the same critical elastic strain (at low temperatures; i.e., temperatures below the Debye temperature of the specimen material). As they occur, they interfere with one another, thereby increasing the stress needed for further deformation.

As an indenter creates an indentation it causes at least three types of finite deformation. It punches material downwards creating approximately circular prismatic dislocation loops. At the surface of the material it pushes material sideways. It causes shear on the planes of maximum shear stress under itself. Therefore, the overall pattern of deformation is very complex, and is reflected

TABLE 2.1

Bond Type	Chin-Gilman Parameter
Metallic	0.0056
Ionic	0.013
Covalent	0.12

in a complicated tangle of dislocation lines, and/or twins, and/or phase transformations. Considerable deformation hardening is associated with moving subsequent dislocations through this tangle.

The dislocations in a tangle can lower their potential energy by aligning themselves to form dipoles and higher multipoles. The stress needed to push subsequent dislocations through a tangle (dipoles and multipoles) is proportional to the elastic shear modulus so it may be expected that the hardnesses of simple metals are proportional to their shear moduli. Figure 2.7 confirms this.

2.2   THE CHIN-GILMAN PARAMETER

For interpreting indentation behavior, a useful parameter is the ratio of the hardness number, H to the shear modulus. For cubic crystals the latter is the elastic constant, C_{44}. This ratio was used by Gilman (1973) and was used more generally by Chin (1975) who showed that it varies systematically with the type of chemical bonding in crystals. It has become known as the Chin-Gilman parameter (H/C_{44}). Some average values for the three main classes of cubic crystals are given in Table 2.1.

C_{44} measures the shear strengths of chemical bonds and the Chin-Gilman parameter indicates how directly they interact with dislocation motions which depends on how localized the bonding is. Thus it is relatively large for covalent bonding which is localized to pairs of atoms (electron pair bonding).

It is quite small for metals where the bonding is spread out over large numbers of atoms; and it has intermediate values for ionic crystals where the overall bonding is delocalized, but local pairs of ions interact strongly. These relationships are discussed in some detail in later chapters.

The difference of the Chin-Gilman parameter for differing types of chemical bonding accounts for the Tabor constant not being three for non-metals.

2.3   WHAT DOES INDENTATION HARDNESS MEASURE?

If slip-line fields do not control plastic indentation, what does? The answer is: not the beginning of the plastic deformation, but the end of it. The end means after deformation hardening has occurred. That is, it is not the initial yield stress, Y_0, that controls indentation, but the limiting yield stress, Y^*. This is

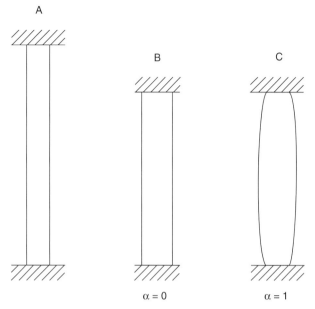

Figure 2.3   Plastic compression of a cylinder. A—Initial configuration. B—compressed cylinder with no frication ($\alpha = 0$) at platen interfaces. C—compressed cylinder with full friction ($\alpha = 1$) at platen interfaces.

reached after substantial plastic deformation has occurred during each incremental advance of the indenter. The fact that the deformation hardening rate is often small has led to confusion about this point in the past.

A rough analog of plastic indentation is plastic compression. Consider the most simple case; compression of a right cylinder between two platens (Figure 2.3A). There are two limiting cases depending on whether the friction coefficient, $\alpha = 0$ (freely slipping platen/specimen interfaces) or $\alpha = 1$ (no slipping at platen/specimen interfaces). The compressive deformation that occurs in the first case (Figure 2.3B) simply consists of a decrease of the length of the cylinder, and an increase in its diameter. On the other hand, in the second case (Figure 2.3C), the diameters at the platens cannot increase so barreling occurs. Thus friction at the loading interface is important. Also, it is clear that compression of the cylinder continues until the specimen has been hardened enough to stop further plastic deformation.

The case of constrained indentation is more complicated than simple compression of a cylinder, but the two phenomena are related. Let the compression cylinder be constrained by surrounding material. This is the case of a flat punch indenting a specimen (Figure 2.4A). Now when the cylinder decreases in length, its diameter is constrained from increasing by the surrounding material. Since volume is conserved during plastic deformation material must flow toward the free surface of the specimen. An amount equal to the volume of the indent (minus the elastic volume change) must mound up on the surface.

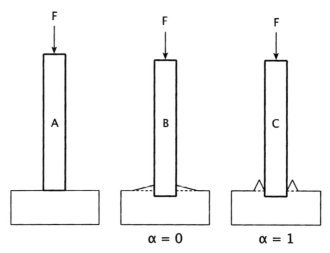

Figure 2.4 Schematic Indentation by cylinder. A—Initial configuration. B—Indentation with zero friction between indenter and specimen. C—Indentation with full friction between indenter and specimen.

The configuration of the mounding depends on whether $\alpha = 0$, or 1. This is illustrated schematically by Figures 2.4B and 2.4C. In practice the mounding is less sharply defined than in these schematic sketches.

The flow toward the surface is caused by the pressure under the indenter. It is analogous to the upward flow around a sphere dropped into a liquid. It is also analogous to inverse extrusion. A model of the flow has been proposed by Brown (2007) in terms of rotational slip. This model reproduces some of the observed behavior, but it is a continuum model and does not define the mechanism of rotational slip.

Next consider a case that is closer to the indenters used for measuring hardness, but is still highly simplified. Figure 2.5 illustrates the incremental penetration of a conical indenter (the conical shape simplifies the practical geometries) into a plastic material. The incremental work done by the load, F on the indenter, is Fdx. The incremental plastic deformation, δ averages about $x/L = \tan\theta$, and the deformed volume is $\pi L^2 \cos\theta\, dx/4$, so with $\theta = 22.5$ the incremental plastic work is: $0.45 Y L^2 dx$ where Y is the stress required for plastic deformation.

During the initial part of the penetration process, the increment of applied work is too large to be balanced by the energy dissipated by the plastic deformation increment. However, the latter continually increases as x increases because of deformation hardening. During each increment, the "yield" stress, Y, starts at the initial yield stress, Y_0, and increases until deformation-hardening increases it up to Y^* (the "saturation" value). Thus the work applied to the indenter during each increment increases until it equals the energy dissipated by the plastic deformation increment. Then no further penetration

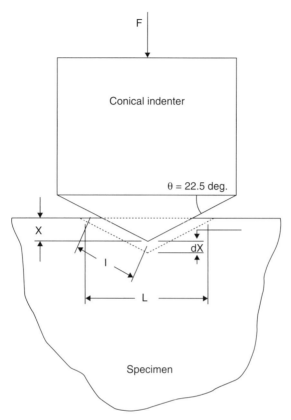

Figure 2.5   Schematic incremental indentation by a conical indenter. See text for explanation of symbols.

occurs. Since the indentation area is $\pi L^2/4$, the hardness number is $4F/\pi L^2$. The contact area of the indenter cone is $(\pi L^2/2)[(L/2)^2 + x^2]$ with $x^2 = [(L/2)\tan\theta]^2$, so the contact area is $1.7L^2$, and the incremental deformed volume is $1.7L^2dx$. Therefore, the dissipated plastic energy is: $\sim 1.7Y{*}L^2dx$. Equating the applied work, Fdx, and the dissipated energy yields: $F = 1.7Y{*}L^2$. Dividing this by the indentation area gives the hardness number: $H = 2.2Y{*}$. Note that the energy is dissipated both by heat generation and by defect storage.

This model is not precise, but does identify some of the factors that are important to indentation. Like the model, the hardness measurement process is not precise. At the micro-hardness level, the projected areas of indentations are measured, but this can only be done with about 10% accuracy. At the nano-indentation level, relative values can determined accurately, but absolute values are probably only about 10% accurate.

Consider the schematic stress-deformation curve of Figure 2.6. Here elastic strain ε dominates until the stress reaches Y_O; then, plastic deformation δ dominates. Note that plastic flow begins as soon as a small stress is applied,

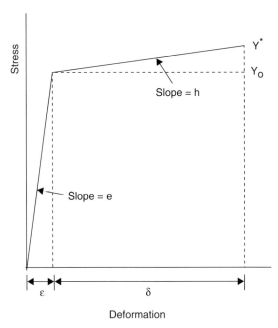

Figure 2.6 Schematic stress-deformation curve with linear deformation-hardening. The elastic, ε and plastic deformations, δ are distinguished.

but it remains relatively small until Y_O is reached. Also, note that plastic deformation is a physically different entity from elastic deformation. Thus, this schematic diagram is very approximate, but it illustrates a general point that must be adjusted when real cases are considered.

The flow stress in Figure 2.6 is given approximately by $Y = Y_O + h\delta$ where h is the deformation-hardening coefficient. It is assumed that the elastic strain is fully recovered in a hardness measurement, so it need not be considered further in this approximate treatment. Then $H = 2.2h\delta$, and since $\delta = 2x/L = \tan\theta = 0.414$; then $H = 0.89h$. Hence $H \simeq h$. That is, the indentation hardness number approximately equals the deformation-hardening coefficient.

This analysis is consistent with the conclusion of Gerk (1977) that the behavior that determines hardness is deformation-hardening; not the "yield" stress. He was one of the first authors to point this out. For other types of materials, it is the maximum stress that the material can bear after deformation (plastic, or that associated with phase transitions in cluding twinning). Hardness is not directly related to the elastic limit, although there is an indirect connection with the offset plastic deformation of metals as demonstrated by Tabor (1951).

Deformation-hardening in the 5–50% deformation range is known to be proportional to either the Young's modulus, Y, or the shear modulus, G, in metals. The Young's modulus depends strongly on the shear modulus since $Y = 2(1+v) G$ where v = Poisson's ratio. For both fcc and bcc pure metals data

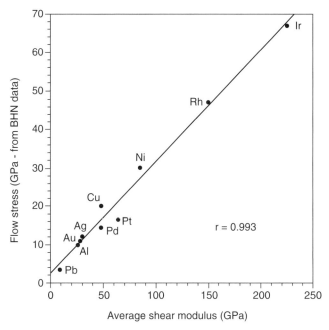

Figure 2.7 Plastic flow stresses of from Brinell spherical indentation-hardnesses *versus* elastic shear moduli. Nominally pure fcc metals at 200 K (Gilman, 1960).

show this dependence, Figures 2.7 and 2.8. The correlation for the fcc metals is clearly much better than that for the bcc metals perhaps because the bcc metals were not sufficiently pure.

Gerk (1977) extended the presentation of Figures 2.7 and 2.8 to a variety of materials. He concluded that the slopes, h of the true stress *versus* true deformation curves equal:

$$h = \beta G \tag{2.1}$$

where G = shear modulus and β = constant. Thus the data are consistent with the idea that the deformation-hardening coefficient is the dominant parameter.

A complication is that the deformation field under an indenter is not homogeneous. It is characterized by local glide bands that form the rosette patterns mentioned earlier (Figure 2.2). This makes the process exceedingly difficult to accurately model using either analytic, or numerical computations.

Gerk showed that Equation (2.1) is followed not only for metals, but also for ionic and covalent crystals if two adjustments are made. For covalent crystals, the temperature must be raised to a level where dislocations glide readily, but below the level where they climb readily. For ionic crystals, G (an average shear modulus) must be adjusted for elastic anisotropy. Thus it becomes:

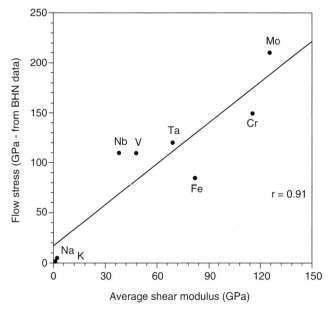

Figure 2.8   Same as Figure 2.6 for bcc metals.

$$G* = G/1 + A \tag{2.2}$$

where   A = anisotropy   factor = $2C_{44}/(C_{11} - C_{12})$ where   the C_{iij} = elastic constants.

2.4   INDENTATION SIZE EFFECT

It is observed that indentations made with low loads on an indenter are smaller than expected from the sizes made with high loads. Thus the apparent hardness of a specimen increases as the indentation size decreases. This is known as the *indentation size effect* (ISE). It has been given a variety of interpretations, but the most simple is that it is associated with friction at the interface between the indenter and the specimen (Li et al., 1993).

The indentation process is driven by the applied load, and resisted by two principal factors: the resistance of the specimen to plastic deformation (and elastic deformation); plus the frictional resistance at the indenter/specimen interface. The ratio of these resistances changes with the size of the indentation because the plastic resistance is proportional to the volume of the indentation, while the frictional resistance is proportional to the surface area of the indentation. Therefore, the ratio varies as the reciprocal indentation size. This interpretation has been tested and found to be valid by Bystrzycki and Varin (1993).

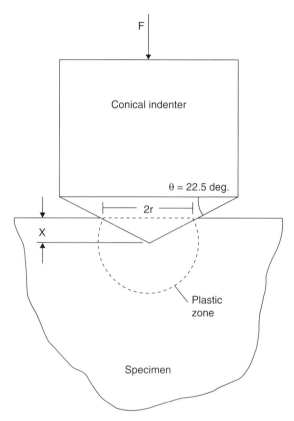

Figure 2.9 Schematic indentation. Similar to Figure 2.5, but emphasizes the plastic zone.

An approximate model of the ISE can be developed with the aid of Figure 2.9. This figure shows a schematic cylindrical indenter with a conical tip being pushed into a specimen. The plastic zone is approximated by a segment of a sphere, and the diameter of the indent is 2r. The yield stress is Y, and the friction coefficient is α.

The friction coefficient is expected to depend on: the normal pressure which is quite high (of order hundreds of kilobars); surface roughness; surface homogeneity; and humidity (or other environmental factors). As a result, α is not known, so a quantitative model is not possible, but the expected qualitative behavior is clear.

The interfacial area is: $\pi r(r^2 + x^2)^{1/2} = 3.36r^2$. The radius of the plastic zone is $r/\cos\theta = 1.08r = R$, so its volume is: $[(4/3)\pi R^3 - (\pi/3)r^2x] = 4.88r^3$.

Neglecting the elastic forces, lumping the geometric factors into a constant, b, and assuming the plastic shear deformation is x/r, yields the plastic resistive force:

$$F_P \approx bYr^2 \tag{2.3}$$

Similarly, for the frictional resistive force, F_F, lumping the geometry factors into a constant, c, and letting α = friction coefficient:

$$F_F \approx c\alpha r \qquad (2.4)$$

Balancing these forces with the driving force, F yields:

$$F \approx F_P + F_F \approx bYr^2 + c\alpha r \qquad (2.5)$$

But the hardness number $H = F/\pi r^2$, so if new constants, B and C are substituted for b and c:

$$H \approx Y(B + C\alpha/r) \qquad (2.6)$$

or:

$$H \approx H_O + CY\alpha/r \qquad (2.7)$$

which is the observed behavior.

For large indentations, $H = BY$; and for small indentations, $H \sim 1/r$. This equation is not expected to be exact because both Y and α may depend on r, but its general form matches the observations.

Also, it should be noted that α can be very large for some interfaces. For example, if some grit gets lodged between the indenter and the specimen, α can be large compared with unity, leading to anomalous hardness values.

Another source of size effects is that dislocations do not behave near surfaces in the same way they do in the interiors of crystals. Perhaps the most fundamental of these is the behavior of screw dislocations emerging from surfaces. Since the maximum shear stress is usually on planes making angles of 45° with surfaces, screw dislocations do not emerge perpendicularly from surfaces. Therefore, there are forces on them tending to shorten them by crossgliding (Gilman, 1961). This causes local deformation-hardening causing the stresses needed to make small indentations larger than those for larger indentations. This mechanism has been verified by Minari and Pichaud (1980) using X-ray topography.

2.5 INDENTATION SIZE (FROM MACRO TO NANO)

If materials were homogeneous, the sizes of indenters and indentations would not matter. However, they are not homogeneous. They are heterogeneous aggregates of various objects and configurations. These include grains, precipitates, interfaces, and ordered arrays of atoms and molecules; as well as dislocation llnes, and distributions of dislocations lines. Therefore, the sizes

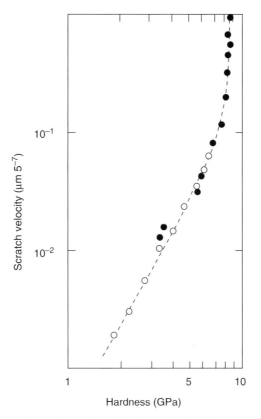

Figure 2.10   Comparison of scratch and indentation hardnesses (after Gerk, 1976).

of indentations relative to the sizes of the heterogeneities affects hardness measurements. In some cases this means that relative hardness values are more reliable than absolute values.

2.6   INDENTATION VS. SCRATCH HARDNESS

Gerk (1976) has shown that indentation and scratch hardnesses are equivalent if the time of indentation and the velocity of scratching are taken into account. Figure 2.10 shows the numerical relationship between the two types of measurement indicating that they are essentially the same.

Correlations between differing measures of hardness are discussed by Mott (1957). A correlation diagram for Mohs scratch hardness and indentation hardness in the case of minerals is given in Figure 2.11.

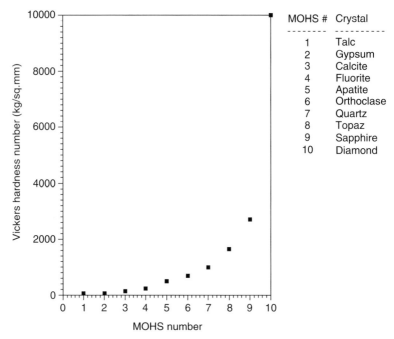

Figure 2.11   Correlation of Mohs (M) and Vickers indentation hardnesses (VHN). Data from Mott (1957). The dependence is roughly exponential. That is: VHN ~ exp (M). Similar to Figure 1.3, but coordinates are linear.

2.7   BLUNT OR SOFT INDENTERS

Standard indenters have sharp points to ensure that indentations always get started in hard materials. However, blunt indenters will usually make indentations providing the load applied to them is greater than some critical value. Furthermore, they need not be harder than the material being indented. This has been demonstrated by Brookes and his colleagues in a series of papers. A good example is their paper on the indentation of MgO (Shaw and Brookes, Mott 1989).

2.8   ANISOTROPY

Since the surfaces of crystals have specific symmetries (usually triangular, square, or tetragonal) and indenters have cylindrical, triangular, square, or tetragonal symmetries, the symmetries rarely match, or are rotationally misaligned. Therefore, the indentations are often anisotropic. Also, the surface symmetries of crystals vary with their orientations relative to the crystallographic axes. A result is that crystals cannot be fully characterized by single hardness numbers.

Knoop indenters are particularly useful for studies of the anisotropies of indentations on surfaces because of their elongated shape which gives an indenter two-fold symmetry. For studies of minerals this is quite useful and has been discussed in some detail by Winchell (1945).

Hardness also depends on which face of a non-cubic crystal is being indented. The difference may be large. For a crystal with tetragonal symmetry the face that is normal to the c-axis can be expected to be different from those that are normal to the a-axes. Similarly the basal faces of hexagonal crystals are different from the prism faces. One extreme case is graphite where the resistance to indentation on the basal plane is very different than the resistance on the prism planes.

Furthermore, crystals whose structures are not centrosymmetric have different hardnesses on opposite sides of a given crystal even though the Miller indices of the surface planes are the same. For example, the hardness of the (0001) plane of ZnS (zinc blende structure) is not the same as that of the (000-1) plane.

2.9   INDENTER AND SPECIMEN SURFACES

Both the indenter and the specimen surfaces should be smooth and homogeneous in order to minimize friction. If the indenter is not smooth, under pressure that is sufficient to cause plastic flow. The specimen will become "embossed" by the indenter, tending to lock the surfaces. This will induce a large effective friction coefficient.

Irregularities of a specimen's surface will result in local deformation with accompanying deformation hardening. This may lead to erroneous hardness numbers, although such errors may be small.

More serious errors may result when the grain-size of a specimen is small compared with the size of an indentation. Then, since all crystals are elastically anisotropic a rigid indenter will produce differing amounts of elastic strain in the grains depending on their orientations. This will create an effective roughening of the surface and increase the friction coefficient. This may result in overestimates of hardnesses. For example, this may underlie reports of nanocrystalline materials being harder than diamond.

REFERENCES

R. W. Armstrong and C. Cm. Wu, "Lattice Misorientation and Displaced Volume for Microhardness Indentations in MgO Crystals", Jour. Amer. Cer. Soc., **61**, 102 (1978).

L. M. Brown, "Slip Circle Constructions for Inhomogeneous Rotational Flow", Maer. Sci. Forum, **550**, 105 (2007).

J. Bystrzycki and R. A. Varin, "The Frictional Component in Microhardness Testing of Intermetallics", Scripta Met., **29**, 605 (1993).

M. M. Chaudhri, "Dislocations and Indentations" in *Dislocations in Solids—Vol. 12*, Edited by F. R. N. Nabarro and J. P. Hirth, p. 447, Elsevier, Amsterdam, Netherlands, (2004).

G. Chin, "Strong and Hard Solids", Trans. Amer. Crystallographic Assoc., **11**, 1 (1975).

A. P. Gerk, "The Relationship of Time-dependent Hardness and Scratch Hardness", J. Phys. D: Appl. Phys., **9**, 179 (1976).

A. P. Gerk, "The Effect of Work-hardening Upon the Hardness of Solids: Minimium Hardness", Jour. Mater. Science, **12**, 735 (1977).

J. J. Gilman, "The Plastic Resistance of Crystals", Australian Jour. Phys., **13**, 327 (1960).

J. J. Gilman, "The Mechanism of Surface Effects in Crystal Plasticity", Phil. Mag., **6** (61), 159 (1961).

J. J. Gilman, "Hardness—A Strength Microprobe", in *The Science of Hardness Testing and Its Research Applications*, Edited by J. H. Westbrook and H. Conrad, p. 51, Amer. Soc. Metals, Metals Park, Ohio, USA (1973).

R. Hill, *The Mathematical Theory of Plasticity*, Clarendon Press, Oxford, UK (1950).

H. Li, A. Ghosh, Y. H. Han, and R. C. Bradt, "The Frictional Component of the Indentation Size Effect in Low Load Hardness Testing", Jour. Mater. Res., **8**(5), 1028 (1993).

F. Minari and B. Pichaud, "Dislocations and Free Surfaces in the Micro-plastic Deformation of F.C.C. Metals", in *Dislocation Modelling of Physical Systems*, Edrs. M. F. Ashby, R. Bullough, C. S. Hartley, and J. P. Hirth, p. 551, Pergamon Press, New York, USA (1980).

B. W. Mott, *Microindentation Hardness Testing*, Btterworths, London (1957).

M. C. Shaw, "The Fundamental Basis of the Hardness Test" in *The Science of Hardness Testing and Its Research Applications*, Edited by J. H. Westbrook and H. Conrad, p. 1, Amer. Soc. Metals, Metals Park, Ohio, USA (1973).

M. P. Shaw, "Dislocations Produced in Magnesium Oxide Crystals Due to Contact Pressures Developed by Softer Cones", Jour. Mater. Sci., **14**, 2727 (1989).

D. Tabor, *The Hardness of Metals*, Clarendon Press, Oxford, UK (1951).

S. N. Timoshenko and J. N. Goodier, *Theory of Elasticity-3rd Edition*, p. 409, McGraw-Hill Book Co., New York (1970).

J. H. Westbrook, "Flow in Rock-salt Structures", General Electric Research Laboratory Report #58-RL-2033 (1958), not published.

H. Winchell, "The Knoop Microhardness Tester as a Mineralogical Tool", Amer. Mineralogist, **30**, 585 (1945).

3 Chemical Bonding

3.1 FORMS OF BONDING

Chemical bonding consists of electrostatic and electrodynamic interactions between valence electrons and positive ions. However, there is more than one category. In small groups of atoms (molecules), pairs of electrons may reside between pairs of ions and electrostatically attract the ions to form bonds (covalent bonding).

Or, electrons from atoms of one type, say A, transfer to atoms of another type, say B, to form two kinds of ions; a positive kind and a negative kind. These, *via* Coulomb's Law of electrostatic attraction, become bonded (ionic bonding).

For larger atomic aggregates, another possibility is that a stable and dense plasma forms, consisting of a swarm of relatively free electrons moving in a background of an equal number of positively charged ions (metallic bonding).

A fourth possibility is electrodynamic bonding. This arises because atoms and molecules are not static, but are dynamically polarizable into dipoles. Each dipole oscillates, sending out an electromagnetic field which interacts with other nearby dipoles causing them to oscillate. As the dipoles exchange electro-magnetic energy (photons), they attract one another (London, 1937).

Hardness is a measure of the resistance of a material to permanent indentation, or scratching. This resistance is determined by the difficulty of shearing one part of the material over another part, and this comes down to the shearing of atoms over one another. To shear one atom relative to others requires the chemical bonds across the shear plane to be rearranged. If the bonding is localized, as it is in the case of electron-pair bonding, individual bonds must actually be broken and then reformed during shearing. If the bonding is not localized as in ionic crystals, and metals, shearing will still disturb it, creating resistance to the shear. The disturbance may consist of changing the local atomic density, or by juxtaposing ions of the same sign to create local electrostatic repulsions. Thus hardness is intimately related to chemical bonding.

The discussion of chemical bonding here is elementary, and is only intended as an outline of the subject. The full subject is very complex (Atkins and

Friedman, 1997), deserving at least a devoted book. However, some knowledge of it is needed to make the atomic basis of hardness comprehensible. The discussion begins with the electronic structures of atoms, then simple molecules, and finally solids. For readers wishing more of the details, an excellent text is that of Oxtoby, Gillis, and Campion (2008).

3.2   ATOMS

Schematically, atoms consist of a positively charged nucleus with a diameter of roughly 10^{-13} cm., surrounded by a cloud of electrons that is roughly 10^{-8} cm. in diameter. Thus the electron cloud is about 10^5 times as large as the nucleus so the latter can be considered to be a positive point-charge. Also, the mass of the nucleus is much larger than that of the electrons, so for most purposes it can be considered to be stationary while the electrons move at high speed around it. For example, take the case of aluminum. The average mass of its nucleus is 27.93 amu. (one atomic mass unit $= 1.66 \times 10^{-24}$ gram). Its atomic number is 13, so there are 13 electrons surrounding its nucleus each weighing 9.11×10^{-28} gram $= 5.5 \times 10^{-4}$ amu. Thus all 13 electrons weigh only about 2.6×10^{-4} as much as the nucleus.

In the lowest energy state (called the ground state) of Al, the electrons occupy various quantum states; written: $1s^2 \, 2s^2 \, 2p^6 \, 3s^2 \, 3p^1$. Here the precursor numbers (called the principal quantum numbers) give the total energy of the state; the letters describe the nature of the state; and the superscripts give the number of electrons in each state (the sum of the superscripts $= 13$). The energy levels of the electrons in Al are:

$$1s = 56.8 \text{ Hartree } (27.2 \, eV.)$$

$$2s = 4.36$$

$$2p = 2.97$$

$$3s = 0.36$$

$$3p = 0.18$$

Thus there are three shells: (1s), (2s + 2p), and (3s + 3p) containing 2, 8, and 3 electrons, respectively. The first two shells are saturated. The third is not, and is the valence shell.

Since Al is only the thirteenth out of more than one hundred elements, it should not be surprising that so far the discussion has only begun to consider the full complexity of the scheme.

In addition to the s and p type states there are d, f, g, ... , states and each of these is subdivided into states having various amounts of angular momentum designated by quantum numbers, l, where:

$$I = 0 \text{ for s-states}$$
$$I = 1 \text{ for p-states}$$
$$I = 2 \text{ for d-states}$$
$$I = 3 \text{ for f-states}$$

. . . .

The quantum number, l, represents the mechanical angular momentum of an electron as it moves around the positive ion core of states with more energy. This angular current generates a magnetic field so the electron has an orbital magnetic moment in addition to its mechanical moment. It has been found experimentally that the angular position of this magnetic moment is spatially quantized. That is, for each value of l, the angular position of the angular momentum vector relative to an external magnetic field vector is quantized. The position of the magnetic vector relative to the mecvhanical angular momentum vector is described by a quantum number, m.

Whatever the other aspects of its state are, an electron itself also has a spin vector. That is, it has a self-state with a quantum number of either +1/2, or −1/2.

In spectroscopic studies all of the above quantum numbers play a role, but for the discussion of hardness only the principal quantum numbers, n; and the angular momentum numbers, l, are usually of importance.

3.3   STATE SYMMETRIES

The s-states have spherical symmetry. The wave functions (probability amplitudes) associated with them depend only on the distance, r from the origin (center of the nucleus). They have no angular dependence. Functionally, they consist of a normalization coefficient, N_i times a radial distribution function. The normalization coefficient ensures that the integral of the probability amplitude from 0 to ∞ equals unity so the probability that the electron of interest is somewhere in the vicinity of the nucleus is unity.

The probability density of an electron with amplitude (wave function) ψ is ψ^2. The s-type (spherical) wave functions, ψ for the first few principal quantum numbers (n = 1, 2, 3 ...) are:

$$\phi(1s) = N_1 e^{-r/2}$$
$$\psi(2s) = N_2(2 - r)e^{-r/2}$$
$$\psi(3s) = N_3(6 - 6r - r^2)e^{-r/2} \tag{3.1}$$

Note that these functions decay exponentially overall but also have nodes at particular values of r.

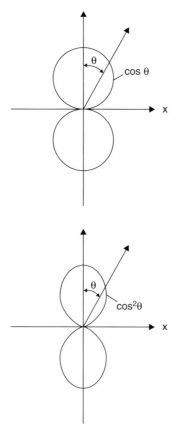

Figure 3.1   Dumbell shape of a p_x-type wave function, say: $\psi_x(r, \theta) = M \cos \theta$, where M = const. (top), and its corresponding charge distribution $\sim \cos^2 \theta$ (bottom).

The p-states have the symmetry of "dumbbells" (Figure 3.1). The axes of the dumbells can have three orientations: p_x, p_y, p_z each directed along one of the Cartesian axes. Consider p_z, for example. Let M = normalization coefficient; then the form is:

$$\psi(2p) = M \cos \theta \tag{3.2}$$

where θ is the angle between r and the z-axis. If all three of the p-states are occupied by electrons, the overall symmetry is spherical. Since the electron density is proportional to the square of the amplitude, or ψ^2, the electron density is proportional to $\cos^2 \theta$ for the p-states.

The d-state probability amplitudes (in two dimensions) are shaped like cloverleafs. For the principal quantum numbers, $n = 1$ and $n = 2$, they lie too close to the nuclei of atoms to interact when the atoms are spaced by s-type wave functions. Only for $n = 3$ and greater, do they extend far enough from their nuclei to cause strong interactions between atoms. Then they play an important role. This is the case in the transition metals and their compounds.

3.4 MOLECULAR BONDING (HYDROGEN)

A hydrogen molecule consists of two atoms with overlapping electron distributions forming an electron-pair bond in the overlap region (Figure 3.2).

A simplified theory based on Heisenberg's Principle and Coulomb's Law will be given here to illustrate the nature of chemical bonding. This approach allows the essence of a bond to be described approximately without solving Schroedinger's equation. The latter can only be solved approximately for the H_2 molecule. The simple theory is possible because much of the content of the Schroedinger equation is contained in Heisenberg's Principle. By combining the latter with general chemical knowledge, and Coulomb's Law of electrostatics, an adequate model can be derived. It is an adaptation of Kimball's method (1950).

In the Heisenberg model the molecule's structure is further simplified. It is envisioned to consist of two overlapping spherical "clouds" of charge with radii, R and volumes, $4\pi R^3/3$. Letting q = electron charge, the charge density, ρ in each cloud is $3q/4\pi R^3$. In addition, there are two positive protons spaced 2r (the bond length) apart. The configuration is shown in Figure 3.3.

Just as the first term in the Schroedinger equation describes the kinetic energy of an electron system, and the second term deals with the potential energy, so are these the two parts of any simple model. Electrostatic forces provide attraction between the atoms, while kinetic energy keeps the system from collapsing by exerting a quantum mechanical "Schroedinger pressure."

Let us start with the kinetic energy, $T = mv^2/2 = p^2/2m$, where $p =$ momentum = mv (m = electron mass, v = velocity). In its exact form, Heisenberg's Principle states that (Born, 1969):

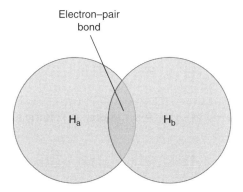

Electron–pair
bond

H_a H_b

Figure 3.2 Overlap of atomic 1s-wavefunctions of H-atoms to form an H_2 molecule with an electron-pair bond.

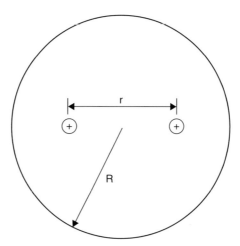

Figure 3.3 Schematic model of a hydrogen molecule with two positive nuclei (separated by distance, r) embedded in a uniform charge cloud with spherical radius, R.

$$\Delta p_x \Delta x = h/4\pi \qquad (3.3)$$

where h = Planck's constant, and Δp_x and Δx are the standard deviations of the x-component of momentum and the x-position. These parameters are described by Gaussian envelopes of carrier waves. The envelopes are centered at the average values of the parameters. The standard deviations describe their spreads.

The radius, R, of each electron cloud is taken to be the root mean square distance of the electron from the center of the cloud. The equation of a sphere of radius, r, is: $x^2 + y^2 + z^2 = r^2$, so on average:

$$<x^2> + <y^2> + <z^2> = R^2$$

but the coordinates are equal, so:

$$<x^2> + <y^2> + <z^2> = R^2/3$$

also, from the symmetry:

$$<x>^2 = <y>^2 = <z>^2 = 0$$

The standard deviation of x is given by:

$$<\Delta x>^2 = <x^2> - <x>^2$$

so:

$$\Delta x = \Delta y = \Delta z = R/\sqrt{3}$$

or, from Heisenberg's Principle:

$$\Delta p_x = \Delta p_y = \Delta p_z = (h/4\pi)(\sqrt{3}/R)$$

since the (+) and (−) components of momentum are equal, the averages of the x, y, and z components = 0; so:

$$<p_x^2> = <p_y^2> = <p_z^2> = (h/4\pi)(3/R^3)$$

Therefore, the kinetic energy is:

$$T = (1/2\,m)\left[<p_x^2> + <p_y^2> + <p_z^2>\right]$$
$$= 9h^2/32\pi^2 mR^2 \qquad (3.4)$$

In atomic units (distance = Bohr radius, $a_0 = (1/m)(h/2\pi q)$; electron charge = 1, and energy = $2m(\pi q/h)^2$) this is:

$$T = (3/2R)^2 \qquad (3.5)$$

which indicates that as a cloud gets smaller, the kinetic energy rapidly increases.

Next, three major contributions to the net electrostatic energy need to be considered. They are the two proton-proton interactions; the four proton-cloud interactions; and the electron cloud-electyron cloud interaction. Start with the potential of a point charge with respect to a charge cloud of radius, R. Figure 3.4 illustrates the division of the electric field into two parts, one when the point charge is outside the spherical cloud $(x > R)$, and the other when it is inside the cloud $(x < R)$. Outside the cloud the field varies as $1/x^2$. Inside, it decreases linearly with x as less and less charge lies closer to the center than it does. These variations are in accordance with Gauss' Theorem.

The field outside the cloud is (cgs. Units):

$$E = q/x^2 \quad (x > R)$$

whereas, inside it is: $E = (q/x^2)(x/R)^3 \quad (x < R)$
the corresponding potential energy is:

$$P = -\int_{\infty}^{x} E\,dx$$

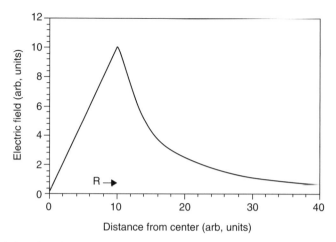

Figure 3.4 Electric field for a positive charge and a sphere of uniform charge.

and:

$$\mathbf{P}\text{(point-point)} = q^2/x \quad (x > R) \tag{3.6}$$

$$\mathbf{P}\text{(point-cloud)} = \left(q^2/R\right)\left(3R^2 - x^2\right)/\left(2R^2\right)$$
$$= \left(q^2/R\right)\left[(3/2) - (1/2)(x/R)^2\right] \quad (x < R) \tag{3.7}$$

To get the potential energy of two overlapping charge clouds, consider the interaction of small elements of each cloud and then form a double integral over each of them. Calling the clouds 1 and 2, and the volume elements within them dv_1 and dv_2, each of which carries a charge equal to the charge density of each cloud times the element's volume, the potential is:

$$\mathbf{P}\text{(cloud-cloud)} = \int\int \{(\rho dv_1)(\rho dv_2)\}/r_{12}$$

where the volume elements lie r_{12} apart. First, this is integrated over the (2) coordinates giving the potential of cloud (1) in the presence of cloud (2):

$$\mathbf{P_2}(1) = \int (\rho dv_2)/r_{12}$$

To get the total potential this is multiplied by the charge elements on cloud (1), and integrated:

$$\mathbf{P}\text{(cloud-cloud)} = \int P_2(1)\rho dv_1$$

We already have an expression for $\mathbf{P_2}$ (Equation 3.7), and for the charge density, so:

$$\mathbf{P}\,(\text{cloud-cloud}) = \int (q/2)\{(3R^2 - x^2)/R^3\}(3q/4\pi R^3)(4\pi x^2 dx)$$
$$= (6/5)(q^2/R) \tag{3.8}$$

The next step is to determine the relationship between r and R. This is done by constructing a force balance. The two protons repulse one another with a force:

$$\text{Repulsion} = 1/(2r)^2 \quad \text{(atomic units)}$$

And each proton is attracted to one of the electron clouds through the point-cloud potential (Equation 3.7). The attractive force is given by the gradient of the point-cloud potential:

$$\text{Attraction} = 2\left(\partial P_{\text{p-c}}/\partial r\right) = 2r/R^3$$

Equating the two forces yields r = R/2.

Putting the terms together to get the total energy, U:

$$U = 2(\text{kinetic}) + 2(\text{proton-proton}) - 4(\text{proton-electron}) + (\text{electron-electron})$$
$$= 2(9/4\,R^2) + 2(1/R) - 4\{[3R^2 - (R/2)^2]/R^3\} + 12/5R \tag{3.9}$$

Setting dU/dR = 0, and solving for R, $R_0 = 15/11 = 0.72$ Angstroms, compared with the experimental value, 0.74 Angstroms.

Substituting R_0 back into Equation (3.9) gives the total energy, U = −121/50 = 33 eV., compared with the experimental value of 32 eV., and the dissociation energy is D = 5.7 eV., compared with the experimental 4.7 eV.

The results of this simple theory are remarkably close to the experimental values, showing that it contains most of the essential ideas in the bonding of the hydrogen molecule. However, the assumption of complete overlap of the electron clouds is not realistic, and it neglects electron spin-spin anti-correlation (i.e., it assumes that the electron spins are anti-parallel). It under-estimates the kinetic energy and over-estimates the electrostatic energy. On the other hand, the origins of the terms in it are quite clear, unlike those in an approximate solution of the Schroedinger equation.

The theory just presented shows how the behavior of electrons leads to bonding in the ground state of a molecule. When dislocations move to produce plastic deformation and hardness indentations, they disrupt such bonds in covalently bonded crystals. Thus bonds become anti-bonds (excited states). This requires that the idea of a hierarchy of states that is observed for atoms be extended to molecules.

In the ground state of a covalent bond, the molecular orbital is occupied by at least one, usually two electrons with anti-parallel spins. This is said to be the HOMO level; that is, the "highest occupied molecular orbital." If the bond is slightly sheared, the kinetic energies of its electrons is not affected, but the electrostatic energies become increased, and a restoring force develops. If the bond is severely sheared, both energies increase and the electrons tend to delocalize in order to reduce their kinetic energy. Thus the electrons go into an anti-bonding state. This is said to be the LUMO level. That is, the "lowest unoccupied molecular orbital." In an aggregate, the molecules become a plasma, or a metal, depending on the density (spacing) of the atoms.

In 1927 Heitler and London showed how the behavior of probability amplitudes (wave functions) leads naturally to bonding and anti-bonding states. Their theory is outlined next. It is approximate, but makes the point. The Schroedinger equation for chemical bonds cannot be solved exactly (the hydrogen molecular ion, H_2^+, is the only exception). The same thing is true for atoms with more than one electron. The electronic structures of atoms with a few electrons can be determined numerically with good accuracy, but not exactly. Thus it cannot be expected that the more complex case will be highly accurate. Instead, as systems of electrons and atoms get complex, theories of them get more approximate. Claims in the literature of five percent theoretical accuracy should be taken with a grain of salt. Therefore, simple, plausible theories tend to be more useful than complex calculations.

A simple argument regarding the nature of bonding comes from Heisenberg (1930). He discusses the resonant transfer of electrons from one atom to another. This is analogous with the transfer of kinetic energy between two equal pendulums coupled by a weak spring. The kinetic energy of one pendulum is gradually transferred to the other, and back again. There are two special points in the motion of the two penduli: one when they are both moving in phase, so the potential energy is at a minimum; and the other when they are moving completely out of phase, so the potential energy is at a maximum. The energy in the minimum case is lower than twice the energy of one of the penduli.

The two electrons in a covalent bond are not confined to the vicinity of either atom, so by moving back and forth between them, the region of their confinement is larger (roughly twice as large). This lowers their kinetic energies according to Heisenberg's Principle, and allows the atomic cores to lie more closely than the diameter of a free atom. Since they lie more closely to the positive ion cores, their electrostatic binding energy increases according to Coulomb's law. On average the electron density is greater between the nuclei than at them.

This can be given a simple quantitative description by means of the Heitler-London molecular orbital theory, but this requires more details than is appropriate in this text. Appendix I provides some of the details for interested readers. Only an outline will be given here.

3.5   COVALENT BONDS

The form of the wave function and its square (the electron density) for the 1s state of a hydrogen atom is shown in Figure 3.5. Consider two hydrogen atoms

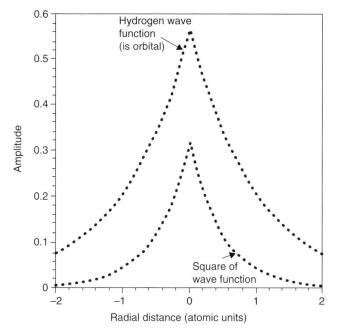

Figure 3.5   Wave function ($\psi = [\pi^{-1/2}]e^{-1/r}$), and its square, for the 1s state of the hydrogen atom.

with wave functions, ϕ_A and ϕ_B. Their electron densities are ϕ_A^2 and ϕ_B^2 as long as they are well separated so their wave functions do not significantly overlap. However, as they become close together, their wave functions will overlap and molecular orbitals $\psi_{a,b}$ will form. Approximately, these will be linear combinations of the atomic orbitals (LCAO's):

$$\psi = \phi_A \pm \phi_B$$

If the individual wave functions are added (i.e., are in phase), the molecular orbital, ψ_b, is bonding. If they are subtracted (i.e., are out-of-phase), the molecular orbital, (ψ_a, is anti-bonding.

The bonding orbital, $\psi_b = \phi_A + \phi_B$, must be squared to get the electron density of the bond:

$$\psi_b^2 = \phi_A^2 + \phi_B^2 + \phi_A \phi_B \tag{3.10}$$

whereas, for the anti-bonding orbital:

$$\psi_a^2 = \phi_A^2 - \phi_B^2 \tag{3.11}$$

The last term in Equation (3.10) results from the overlap of the wave functions, which results in an accumulation of charge between the atoms, forming

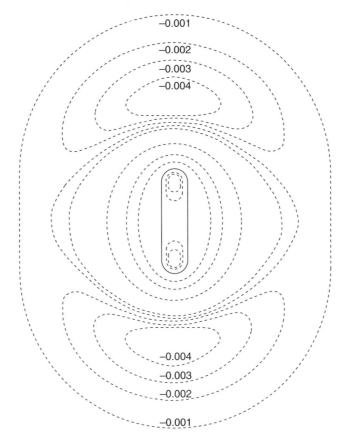

Figure 3.6   Energy density map for the difference between a hydrogen molecule and two hydrogen atoms. The solid lines indicate increased density and are marked with labels indicating atomic units. The dashed lines indicate decreased density. Two dots indicate the positions of the hydrogen atoms in the molecule. They lie 0.74 Angstrom units apart. After Bader and Henneker (1965).

a covalent bond providing the spins of the two electrons that occupy the orbital are anti-parallel. Equation 3.11 indicates that no accumulation between the atoms occurs in the case of anti-bonding.

For the bonding case, Figure 3.6 (from Bader and Henneker, 1965) is a map of the electron difference density in the plane of the bond. It shows the density difference between the molecule and the two atoms. In the region of the center of the bond the density is increased by about 0.08 atomic units as a result of the overlap term in Equation 3.10. Just outside the region of the molecule the difference is diminished from what it would be at the same distance from an isolated hydrogen atom by more than 0.004 units.

At the equilibrium bond length (0.74 Ang.), the energy of the H_2 system in the bond state is at a minimum of $-4.7\,eV$ relative to two separated H atoms.

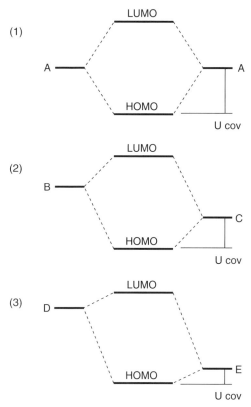

Figure 3.7 Diagrams of the effect of overlapping of the atomic wavefunctions (electron clouds) on the energy levels in molecules. The increased electron density in the bond region decreases the electrostatic energy, while it increases the kinetic energies of the electrons, thereby limiting the closeness of approach of the protons. The figure shows three cases: (1) two identical atoms, A and A with their energy levels splitting into a lower bonding level (HOMO—highest occupied molecular orbital), and a higher anti-bonding level (LUMO—lowest unoccupied molecular orbital). The bonding energy, U_{cov} is the difference between the atomic levels and the HOMO level; (2) two atoms, B and C with moderately different energy levels; (3) two atoms, D and E with a large difference in their energy levels.

The anti-bonding state lies about the same amount above the level of the separated atoms, or +4.7 eV. Thus the energies of the valence electrons on the H-atoms split into increased and deceased values in the H_2 molecule. This is shown schematically in Figure 3.7 and introduces the next topic, the variation of the energies with the separation distance.

Numerical solutions of the Heitler-London, or of density functional equations, show how energies depend on separation distance, but it is more instructive to consider semiempirical equations such as the Morse potential, or especially, the very simple Rydberg equation which has been shown to apply

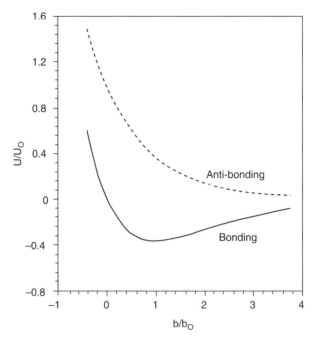

Figure 3.8 Dependence of bonding and anti-bonding energies on bond length. Normalized units.

nearly universally to chemical bonds and solids (Universal Bond Energy Relation, Sutton, 1994). The Rydberg equation has the non-dimensional form:

$$U(b)/U_0 = -(b/b_0)\exp(-b/b_0) \tag{3.12}$$

where b = bond length, b_0 = equilibrium bond length, U_0 is the dissociation energy, and $U(b)$ is the energy. This gives the bonding energy curve. The anti-bonding energy curve is obtained if $[-(b/b_0)]$ is omitted. These curves are shown schematically in Figure 3.8.

If the bonding wave functions have different energies, the bond energy is decreased. Figure 3.9 illustrates this. It compares the energy diagram for molecule formation when two identical atoms, A and A combine with the case for two different atoms, C and D. The covalent bond energy is less in the latter case. However, an ionic bond of significant strength may form.

For strong covalent bonds, then, large wave-function overlaps are desirable (high valence-electron densities), and have similar valence-electron energies. Thus the C-C covalent bond is strong, whereas in LiF—with quite different valence-electron energy levels—the covalent bond is weak, but the ionic bonding is strong. Note that the LUMO-HOMO energy gap is larger when the atomic valence-electron energy levels are different.

However, the energy gap densities are in opposite rank. Using the bond energies, 83 for C-C molecules and 136 kcal/mole for LiF molecules, and the

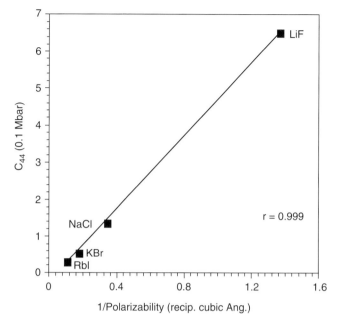

Figure 3.9 C_{44} elastic moduli *vs.* reciprocal polarizabilities for prototype alkali halide crystals.

solid-state molecular volumes, 3.65 for C-C and 8.12 Å^3 for LiF, the bond energy densities are 23 and 17 kcal/mole-cu.Ang., respectively. Since elastic stiffness has the dimensions of energy-per-unit volume, this tends to connect the mechanical behavior of molecules with the electrostatic part of the mechanics of solids. The electrodynamic bonding forces must be dealt with separately.

3.6   BONDING IN SOLIDS

3.6.1   Ionic Bonding

Return to the case of LiF. Lithium ionizes readily, but has little affinity for electrons (I = ionization energy = 5.4 eV and A = electron affinity $\cong 0$ eV.). On the other hand, fluorine is difficult to ionize, but has considerable electron affinity (I = 17.4 eV. and A = −3.6 eV.). Thus, when Li and F atoms are close neighbors, electrons can transfer to make Li$^+$ and F$^-$. These then attract electrostatically until compression of their ion-cores prevent them from contracting further. In a solid crystal, there are both attractive +/− pairs, and repulsive (+/+ as well as −/−) pairs. However, for large arrays, there is a net attraction. This can be shown most simply by examining a linear chain of +q, and −q charges (Kittel, 1966).

   In a linear +/− chain with the charges separated by a distance, d picks a central charge. Then, the electrostatic energy of interaction of the central charge with its nearest neighbor on the right is—letting $q^2 = 1$ $-1/d$. The next-nearest neighbor and the central charge interact with an energy $+1/2d$, and the next-next-nearest neighbor $-1/3d$, and so on. The interactions of the nearest neighbors beyond the first one are neglected because they nearly cancel. Multiplying by 2 because there is a left-hand side of the array, the total energy is a constant (M = Madelung's constant) divided by d, or:

$$M/d = 2[1/d - 1/2d + 1/3d - 1/4d + \ldots\ldots]$$

so:

$$M = 2[1 - 1/2 + 1/3 - 1/4 + \ldots\ldots]$$

but:

$$\ln(1 + x) = x - x^2/2 + x^3/3 - x^4 + \ldots\ldots$$

finally:

$$M = 2\ln 2 = 1.39$$

The calculation of M for a three-dimensional array is much more complicated, and depends on the structure of the array. For the particular case of the face-centered-cubic NaCl crystal structure, its value is M = 1.747, whereas, for the body-centered-cubic CsCl structure, it is M = 1.763.

In ionic crystals with d = nearest neighbor distance, the ions repulse each other strongly when d becomes smaller than the equilibrium value d_0. This can be described by an inverse power function, $+1/d^n$, where n is a power of order, 9. As for the electrostatic attractions, these repulsions must be summed over the N molecules of the crystal structure, yielding another constant, D. The energy, ϕ per molecule (ion pair) is then:

$$\phi = -Mq^2/d + D/d^n$$

forming the derivative of this:

$$\partial\phi/\partial d = Mq^2/d^2 - nD/d^{n+1} = 0 \text{ when } d = d_0$$

which may be used to eliminate D from the previous equation, and obtain an expression for the equilibrium energy, U_0:

$$U_0 = N\phi = -NMq^2/d_0[1 - (1/n)] \tag{3.13}$$

For sodium chloride, since there are eight molecules per unit cell, and the cell volume is a_0^3 where a_0 is the lattice parameter, the energy per unit volume becomes:

$$u = U_0/V = -14q^2/d_0^4[1-(1/n)] \tag{3.14}$$

Note that with $n \gg 1$, the bonding energy is mostly electrostatic attraction.

3.6.2 Metallic Bonding

Metallic bonding is an enigma because there is no single characteristic that defines a metal. More than 80 percent of the chemical elements are metals as defined by their metallic conductivities. On the other hand, metals were originally valued for their strength, particularly their ductilitiy, rather than their conductivity. However, several metals, and particularly their compounds, are brittle; but good conductors. Also, the polymers that have metallic conductivity are clearly covalently, not metallically, bound. Furthermore, simple metallic bonding is spherically symmetric, but many metals are anisotropic such as, zinc. All this means that metallic bonding is not consistently simple.

Only the most simple form of metallic bonding will be considered here. In its simple form a metal is a dense plasma of nearly free electrons and positive ions. The ions are condensed into close-packed 3-D face-centered arrays. Metallic bonding results from a balance between attractive potential energy and repulsive kinetic energy.

Since each atom in a f.c.c. array of purely metallic atoms is the same as every other atom (except at the surface), only a representative positive ion needs to be considered. Let it interact with a spherical portion (radius $= R$) of the electron gas which has a density of one electron per ion. This is called the "jellium" model.

The potential energy is the sum of Equation 3.7, and because there is only one valence electron one-half of Equation 3.8. That is $(-3/2 + 3/5)$ $q^2/R = -9/10q^2/R$.

The kinetic energy in this case is the Fermi energy $= +(1/35.6)(h^2/mR^2)$. So, the total energy of the equivalent atom has the form:

$$U(R) = -\alpha/R + \beta/R^2 \tag{3.15}$$

where $\alpha = 9q^2/10$, and $\beta = (1/35.6)(h^2/m)$. Differentiating Equation 3.15 with respect to R, and setting the derivative $= 0$, the equilibrium value of R is found: $R_0 = 2\beta/\alpha = 1.3$ Ang., and the equilibrium energy is: $U_0 = -\alpha^2/4\beta = -5.03$ eV. (Kittel, 1966).

Jellium theory has shortcomings as it stands as shown in Equation 3.15 for at least two reasons. First, because it has no shear stiffness. Second, because it is not stable. For $R < 2$, its surface energy is negative, so the electron gas is expected to expand indefinitely. Shore and Rose (1991) have proposed a method for stabilizing it, but their method does not give it shear stiffness. Both shortcomings can be solved by recalling that metals become dielectrics at very high frequencies (ultra-violet light frequencies). A simple theory of the critical frequencies was given by Zener (1933). These frequencies are those of oscillations in plasmas, and are now called plasmons.

Within a jellium atom, the electron frequency is of order 10^{17}/sec. compared with the plasmon frequency for jellium (1.1×10^{16}/sec.) so an isolated jellium atom behaves as a dielectric. However, the valence electron screens any electric field caused by polarization. The screening length (Thomas-Fermi) is 0.47 Ang., or 0.36 of the radius of the jellium atom. Thus the field of the positive ion is reduced by about 30% at R.

Zero-point oscillations will cause fluctuations of the positive and negative centers of charge in jellium atoms. Therefore, between two of them there will be dipole-dipole interactions with bonding energy given by London's theory. Using an estimate for the interaction energy of one-pair of atoms from Kittel (1966, p. 84), and correcting for the Thomas-Fermi screening factor for each atom = $0.64 \times 0.64 = 0.41$:

$$U_{dd}(R) \cong -0.41[2q^2/R] = -9\,eV.$$

which is ample stabilization and probably an overestimate.

This method of stabilizing jellium also generates a mechanism for providing shear stiffness. If a jellium atom is polarized, an included circular plane passing through its center becomes an ellipse. The same is true of a solid sphere that is sheared mechanically. If the shape changes are the same the forces must be the same. The electrical force is qE where q is the charge and E is the electric field. This induces an electric dipole moment: qx where x is the separation distance of the positive, and negative, centers of charge; and equals αE where α is the polarizability. The electrical force becomes: $q^2 x/\alpha$. The mechanical shear strain is x/R, so the strain energy is $U_m = (2\pi/3)\,GRx^2$ and the mechanical force is the gradient of this: $dU_m/dx = (4\pi/3)GRx$. Equating the forces, and solving for G yields (Gilman, 1997):

$$G = (3/4\pi)q^2/\alpha R \qquad (3.16)$$

But α approximately equals an atomic volume, $(4\pi/3)\,R^3$, so $G \cong (3q/4\pi)^2(1/R)^4$. Note that Equation 3.16 shows that the shear stiffness is inversely proportional to the polarizability. This is confirmed in Figure 3.8 and is an important aspect for understanding hardness.

3.6.3 Covalent Crystals

If electron-pair, or covalent, bonding is periodic in two or three dimensions, crystals result. The most important case is the carbon-carbon bond. If it is extended periodically in two-dimensions the result is graphite; in three-dimensions it is diamond. Other elements that form electron-pair bonds are Si, Ge, and α-Sn. Some binary compounds are AlP (isoelectronic with Si),

GaAs (isoelectronic with Ge), InSb (isoelectronic with α-Sn), and other III-V as well as II-Vi compounds.

The gaps in the bonding energy spectra of the corresponding molecules (diatoms in the case of the elements) get carried over into the crystals where they are called band-gaps. They determine the stabilities of the bonds, and the crystals. As Chapter 5 discusses, they also determine hardnesses.

3.7 ELECTRODYNAMIC BONDING

The electrodynamic forces proposed for stabilizing jellium provide the principal type of bonding in molecular crystals such as solid methane, rare gas crystals, solid anthracene, and the like. These forces also form the inter-chain bonding of long-chain molecules in polymeric materials (the intra-molecular bonding within the chains is usually covalent).

To obtain a clear understanding of electrodynamic bonding, start with the field of a static electric dipole. Then, let the dipole oscillate so it emits electromagnetic waves (photons). Consider what happens when the emitted field envelopes another dipole (London, 1937). Finally, determine the factors that convert neutral molecules into dipoles (that is, their polarizabilities).

Any given molecule has two centers of charge, one associated with the positive nuclei of the ion cores, the other associated with the negative valence electrons. For a spherically symmetric molecule (and others) these centers are coincident. When an electric field is applied to such a molecule (or to a solid containing such molecules), the centers of charge separate by some distance, x forming an electric dipole with a moment, $\mu = qx$ where q is the amount of charge associated with each center. μ and x are determined by the polarizability which will be considered later.

The electric field created by a dipole is found using the sketch of Figure 3.9

The electrostatic potential at a point P distant from a dipole of moment, $\mu = qx$ is:

$$P = q(-1/x_- + 1/x_+) = q[(x_- x_+)/(x_- + x_+)] \cong q(x/r^2) \cong \mu/r^2 \qquad (3.17)$$

The electric field is the gradient of the potential, or:

$$E \cong -2\mu/r^3 \qquad (3.18)$$

This simple derivation omits the angular dependence of the field which varies as the cosine of the angle between the dipole axis (the moment vector) and the distance expressed as a vector, **r**. Therefore, the field is a maximum along the axis of the dipole. Equation 3.18 makes the point that the dipole field decreases rapidly with distance. The units here are electrostatic (CGS) for simplicity.

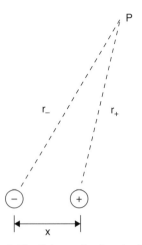

Figure 3.10   Schematic electric dipole.

Now, if x oscillates, a dipole emits an electromagnetic wave (Kittel, 1971) (Figure 3.10). Or, if an electromagnetic wave impinges on a dipole, x will begin to oscillate. Furthermore, dipoles oscillate at all temperatures because of the zero-point energies of quantum oscillators. Therefore, pairs of atomic (or molecular oscillators) always exchange photons. If the exchanges are in phase, this lowers the energy of the pair so there is an attractive force between the two oscillators of the pair. Two identical oscillators will be considered next.

The interaction of two harmonic oscillators with spring constants = k, and masses = m, separated by a distance = r is to be considered. Each atomic (molecular) oscillator has a frequency, $v = (1/2\pi)\sqrt{(k/m)}$. Their orientations are assumed to be collinear (Figure 3.11).

The quantum mechanical energies, U of the oscillators are:

$$U = (n + 1/2)hv$$

so their zero-point energies (n = 0) are hv/2. The energy of the system for very large separations is then: $U(\infty) = hv$. However, for intermediate distances, *circa.* r = c/v, where c = light speed, retardation effects become important, and theory initiated by Casimir must be applied.

For relatively small separations, Figure 3.10 indicates that there are four electrostatic interactions; two positive ones, and two negative ones, so the potential energy of the system is:

$$q^2\{[1/R]+[1/(R+2x)]-2[1/(R+x)]\} \tag{3.19}$$

clearing the fractions, and neglecting small terms yields:

$$2(q)^2/R^3 = 2q^2/R^3 \tag{3.20}$$

Figure 3.11   Two interacting collinear dipoles.

This interaction splits the force constants, k of the individual oscillators into a larger one, $k_+ = k + 2q^2/R^5$, and a smaller one, $k_- = k - 2q^2/R^5$. The oscillators then have modified frequencies:

$$v_+ = v\sqrt{[1+(2q^2)/(kR^5)]} \quad \text{and:} \quad v_- = v\sqrt{[1-(2q^2)/(kR^5)]} \qquad (3.21)$$

so the lowest energy is now the sum of the zero-point energies for the interacting oscillators:

$$U(R) = h/2(v_+ + v_-) = hv[1 - q^4/(2k^2R^6)] \qquad (3.22)$$

and the difference between this and $U(\infty)$ is the interaction energy:

$$U(R) = -(q^4 hv)/(2k^2 R^6) \qquad (3.23)$$

This can be put into more useful form by noting:

1. polarizability = α = qx/E = q^2/k
2. hv = 2 × chemical hardness = 2η
3. in 3D vs. 1D the constant, 1/2 ⇒ ¾

Then, Equation 3.23 becomes:

$$U(R) = -(3/2)\eta\alpha^2/R^6 \qquad (3.24)$$

which is essentially the same as London's equation, the chemical hardness, η defined to be one-half the electronic excitation energy. That is half the LUMO-HMO gap of a molecule, or half the band gap of a covalent crystal. Note that the presence of h (Planck's constant) in Equation 3.23 shows that this bonding is a quantum mechanical phenomenon. It results from the zero-point vibrations of quantum oscillators.

3.8   POLARIZABILITY

Polymeric materials became increasingly important during the second two-thirds of the twentieth century. They consist of neutral molecules held together by London forces. Equation 3.24 indicates that the most important factor (other than molecular size) is polarizability. Experimentally, this is determined from refractive index measurements. The factors that determine it are revealed by quantum mechanics since it involves distortions of atoms and molecules.

Consider a spherically symmetric molecule with superimposed centers of positive and negative charge, q. Apply an electric field (perhaps from a light wave), E(x). This will tend to separate the charge centers by an amount, x; polarizing the molecule, and making it an electric dipole. The dipole moment will be qx = μ which equals αE where α is the polarizability.

From the viewpoint of quantum mechanics, the polarization process cannot be continuous, but must involve a quantized transition from one state to another. Also, the transition must involve a change in the shape of the initial spherical charge distribution to an elongated shape (ellipsoidal). Thus an s-type wave function must become a p-type (or higher order) function. This requires an excitation energy; call it Δ. Straightforward perturbation theory, applied to the Schroedinger aquation, then yields a simple expression for the polarizability (Atkins and Friedman, 1997):

$$\alpha = (2/\pi)[qh/2\pi\Delta]^2 \tag{3.25}$$

This equation indicates that polarizability is proportional to the inverse square of the excitation energy. Therefore, atoms, molecules, and solids with small values of Δ are easily polarized. That is, they are chemically and mechanically soft. The gaps in their bonding energy spectra are small. Since they absorb light easily, they tend to be colored. If Δ lies in a narrow band as in a dye, the coloration is bright and saturated. If it lies in broad band as in adhesive polymers, it may be a muddy brown.

In dielectric materials (oxides. semiconductors, halides, polymers, and he like), polarizability correlates with hardness. For metals, this is not the case. However, the frequencies of the collective polarizations known as *plasmons* are related to mechanical hardness.

REFERENCES

P. W. Atkins and R. S. Friedman, *Molecular Quantum Mechanics—3rd Edition*, Oxford University Press, Oxford, UK (1997).

R. F. Bader and W. H. Henneker, "The Ionic Bond," Jour. Amer. Chem. Soc., **87**, #14, 3064 (1965).

M. Born, *Atomic Physics—8th Edition*, p. 433, Dover Publications, New York, USA (1969).

J. J. Gilman, "Bulk Stiffnesses of Metals," Mater. Sci. & Eng., **7**, 357 (1971).

J. J. Gilman, "Chemical and Physical Hardness," Mat. Res. Innovat., **1**, 71 (1997).

W. Heisenberg, *The Physical Principles of the Quantum Theory*, Translated by C. Ekart and F. C. Hoyt, p. 142ff, Dover Publications, New York, USA (1930).

G. E. Kimball, Lectures at Columbia University, New York (1950).

C. Kittel, *Introduction to Solid State Physics*, 3rd Edition, p. 245, J. Wiley & Sons, New York (1966); also, *5th Edition*, p. 78 (1971).

F. London, "The General Theory of Molecular Forces," Trans. Farad, Soc., **33**, 8 (1937).

D. W. Oxtoby, H. P. Gillis, and A. Campion, *Principles of Modern Chemistry–Sixth Edition*, Thompson Brooks/Cole, Belmont, CA, USA (2008).

H. B. Shore and J. H. Rose, "Theory of Ideal Metals," Phys. Rev. Lett., 2519 (1991).

A. P. Sutton, *Elecronic Structure of Materials*, Clarendon Press, Oxford, UK, (1994).

C. Zener, Nature. **132**, 968 (1933).

4 Plastic Deformation

4.1 INTRODUCTION

Hardness indentations are a result of plastic, rather than elastic, deformation, so some discussion of the mechanisms by which this occurs is in order, especially since the traditional literature of the subject is confused about its fundamental nature. This confusion seems to have arisen because it was considered to be a continuous process for a great many years, and because some metals behave plastically on the macroscopic scale in a nearly time-independent fashion. During the twentieth century, it became well established that plastic deformation is fundamentally discontinuous (quantized), and a time-dependent flow process.

Clarity requires that a distinction be made between elastic strain and plastic deformation. They both have units of length/length, but they are physically different entities. Elastic strain is recoverable (conservative); plastic deformation is not (non-conservative). At a dislocation core, where atoms exchange places via shear, the plastic displacement gradient is a maximum as it passes from zero some distance ahead of the core, up to the maximum, and then back to zero some distance back of the core. In crystals with distinct bonds, the gradient becomes indefinite (infinite) at the core center.

Plastic deformation is a transport process in which elements of displacement are moved by a shear stress from one position to another. Unlike the case of elastic deformation, these displacements are irreversible. Therefore, they do not have potential energy (elastic strain energy) associated with them. Thus, although the deformation associated with them is often called "plastic strain," it is a fundamentally different entity than an elastic strain. In this book, therefore, it will be called plastic *deformation*, and the word *strain* will be reserved for elastic deformation.

Although elastic strain and plastic deformation are expressed as numbers and have the same units (length/length), since they are physically different entities, they cannot be mixed in arithmetic operations. That is, mixtures of them cannot be added, subtracted, multiplied, or divided. Therefore, separated equations should describe them. Constitutive equations that combine them into a single equation are physically meaningless. A consequence is that elastic

Chemistry and Physics of Mechanical Hardness, by John J. Gilman
Copyright © 2009 John Wiley & Sons, Inc.

strain is conveniently described in quasi-static terms, whereas plastic deformation is conveniently described by integrating a rate equation.

Elastic strain results from a "concerted" process (at scales greater than atomic dimensions), whereas plastic deformation results from a "disconcerted" one.

In textbooks, plastic deformation is often described as a two-dimensional process. However, it is intrinsically three-dimensional, and cannot be adequately described in terms of two-dimensions. Hardness indentation is a case in point. For many years this process was described in terms of two-dimensional slip-line fields (Tabor, 1951). This approach, developed by Hill (1950) and others, indicated that the hardness number should be about three times the yield stress. Various shortcomings of this theory were discussed by Shaw (1973). He showed that the experimental flow pattern under a spherical indenter bears little resemblance to the prediction of slip-line theory. He attributes this discrepancy to the neglect of elastic strains in slip-line theory. However, the cause of the discrepancy has a different source as will be discussed here. Slip-lines arise from *deformation-softening* which is related to the principal mechanism of dislocation multiplication; a three-dimensional process. The plastic zone determined by Shaw, and his colleagues is determined by *strain-hardening*. This is a good example of the confusion that results from inadequate understanding of the physics of a process such as plasticity.

4.2   DISLOCATION MOVEMENT

Inelastic deformation rarely, if ever, occurs via homogeneous shear. Dislocations form heterogeneously, and by moving, they transport displacement, thereby causing plastic deformation in proportion to their motion. The amount of displacement they carry in crystals is quantized and called the Burgers displacement. In non-crystalline materials (e.g., glasses) the displacement is somewhat variable, but the variations are small. In granular materials (e.g., sand) the displacement approximates the average size of a granule.

Figure 4.1 schematically shows the definition of a dislocation. Here, on a glide plane, a unit of shear has occurred locally over part of the area of the plane. The boundary of the local area is the dislocation line; in the figure it is one quarter of a circle that enters the crystal on the left front face and emerges on the right front face. The unit of glide (shear) is a translation vector of the crystal structure, called the Burgers vector. It can make any angle with the tangent vector of the line, but there are two special cases where the dislocation's properties are distinctly different.

When the two vectors are parallel, the crystal planes perpendicular to the line form a helix, and the dislocation is said to be of the screw type. In a nearly isotropic crystal structure, the dislocation is no longer associated with a distinct glide plane. It has nearly cylindrical symmetry, so in the case of the figure it can move either vertically or horizontally with equal ease.

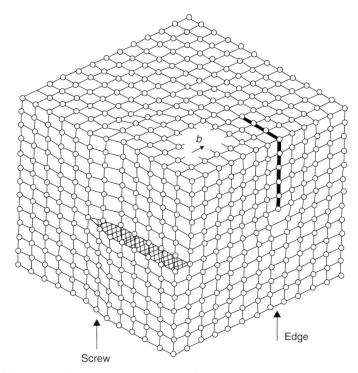

Figure 4.1 Schematic dislocation line a simple cubic crystal structure. The llne enters the crystal at the center of the left-front face. It emerges at the center of the right-front face. The shortest translation vector of the structure is the Burgers Vector, **b**. The line bounds the glided area of the glide plane (100) from the unglided area.

When the line is perpendicular to the displacement (Burgers) vector, the configuration on the right front face in the figure arises. It is characterized by the bold extra half-plane of atoms whose edge ends at the dislocation line; so it is said to be of the edge type. The symmetry at the dislocation line is low, and now gliding motion is limited to the glide plane. It can move off the glide plane (climb) only if atomic diffusion occurs. The material below the edge is extended, while that above is compressed.

Being the edge of a sheared area, a dislocation is a line, but does not, in general, lie on one plane, so its motion is usually three-dimensional. Since shear has two signs (plus and minus) so do dislocations; and dislocations of like signs repel, while those of opposite signs attract. In some structures, the Burgers vector is an axial vector, so plus shear differs from minus shear (like a ratchet).

A key feature of the motion of dislocation lines is that the motion is rarely concerted. One consequence is that the lines tend not to be straight, or smoothly curved. They contain perturbations ranging from small curvatures to cusps, and kinks. In covalent crystals where there are distinct bonds between the top

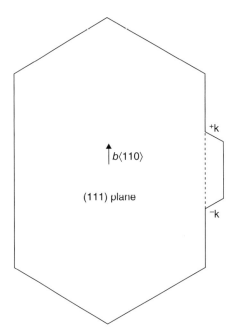

Figure 4.2 Quasi-hexagonal dislocation loop lying on the (111) glide plane of the diamond crystal structure. The <110> Burgers vector is indicated. A segment, displaced by one atomic plane, with a pair of kinks, is shown a the right-hand screw orientation of the loop. As the kinks move apart along the screw dislocation, more of it moves to the right.

and bottom of the glide plane, the lines contain sharp kinks that lie at the chemical bonds (Figure 4.2). As these kinks move along a line with some velocity, v_k, the line moves perpendicular to itself at a velocity, $v_d = c_k v_k$ where c_k is the atomic concentration of kinks. Thus the dislocation velocity, v_d in this case, is determined by the kink velocity.

The energy needed to move a kink is approximately equal to the energy needed to form a pair of them. This activation energy is that of one of the electron pair bonds across the glide plane; since two electrons are involved, it is twice the magnitude of the energy band-gap (Figure 4.3). Note that there is little scatter of the points for the Group IV elements (Ge, Si, SiC, and C). There is some scatter for the isoelectronic III-V compound crystals (InSb, and GaSb). And, much scatter for the other III-V compound crystals (InAs,GaSb, InP, and GaP). The scatter may result from interactions with the environmental atmosphere. In spite of the scatter, the trend seems clear. Note that these activation energies are not directly related to indentation hardnesses because they have been measured at temperatures above the crystal Debye temperatures. The data of Figure 4.3 strongly indicate that bond energies control dislocation motion in covalent crystals.

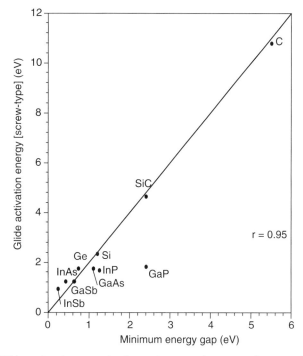

Figure 4.3   Glide activation energies for various covalent crystals *versus* their minimum energy gaps. The correlation coefficient is 0.95. Without the point for GaP, it would be much higher.

When there are no distinct bonds crossing a glide plane, there are no distinct kinks. This is the case for pure simple metals, for pure ionic crystals, and for molecular crystals. However, the local region of a dislocation's core still controls the mobility in a pure material because this is where the deformation rate is greatest (Gilman, 1968).

4.3   IMPORTANCE OF SYMMETRY

If a dislocation line lies parallel to the x-axis of an xy-plane, and is kinked, the kink lies parallel to the y-axis. Therefore, if the line is of edge character, the kink is of screw character. If the line is of screw character, the kink is of edge character. In either case, the displacement gradient is indefinite at the center of the kink. This means that whatever symmetry exists in the undislocated crystal, structure is destroyed at a kink.

A tenet of the quantum theory of the chemical bond is that wave functions on an atom (A) do not form bonds with the wave functions on another atom (B) unless they have the same symmetry (Coulson, 1952). This is because overlapping wave functions interfere destructively unless they have the same symmetry. Therefore, at the center of a kink, where the symmetry is quite different from the normal crystal, chemical bonding becomes essentially non-existent. Thus kinks in covalent crystals lie in potential wells of depth equal to a bond energy.

The discussion just above does not hold for the other types of bonding. Thus, in metallic and ionic and crystals, as well as dispersive crystals, cohesion is not atomically localized, so there are no "bonds" of significant strength. In simple metals, the difference in cohesive energy between the f.c.c. structure and the b.c.c. structure is of order milli-electron-volts, whereas the overall metallic cohesive energy is of order electron-volts. Thus, the local atomic arrangement has only a minute effect on the cohesion. In these cases other factors determine the resistance of the material to inelastic deformation. In metals, it is strain-hardening. In ionic crystals, it is the repulsion of ions having the same charge. In molecular crystals, it is polarizability. In some compounds, it is the heat of formation. And so on. Each type of material must be considered individually. There is no way to generate a general theory because the bonding type varies from one material to another. Nevertheless a simple theory will be provided for each bonding type.

4.4   LOCAL INELASTIC SHEARING OF ATOMS

Inelastic shearing of atoms relative to one another is the mechanism that determines hardness. The shearing is localized at dislocation lines and at kinks along these lines. The kinks are very sharp in covalent crystals where they encompass only individual chemical bonds. On the other hand, in metal crystals they are often very extended. In metallic glasses they are localized in configurations that have a variety of shapes. In ionic crystals the kinks are localized in order to minimize the electrostatic energy.

A generic process of inelastic shearing determines hardness for most materials. Metals are an exception. Two factors play key roles. One is the applied work (energy) required for shearing, and the other is the energy barrier that resists the shearing. These lead to a generic parameter; (energy/volume), called here the "bond modulus." The barrier to shearing may be a chemical bond in covalent crystals, or the heat of formation of a compound, or the heat of mixing in a solid solution. In simple metals, it is other dislocations (strain-hardening).

An antecedent of the bond modulus is the chemical hardness of Pearson (1997) which measures the stabilities of molecules. Also, bond moduli are proportional to the physical hardnesses of Yang, Parr, and Uytterhoeven (1987) which they proposed for minerals.

The bond modulus is a quite simple parameter. It is the ratio of an atomic (or molecular) energy, and an atomic (or molecular) volume. Essentially all of the volumes of interest are known from crystallography, and the energies are either known from physical chemistry, or can be estimated. There is no need for extensive numerical calculations in most cases.

4.5   DISLOCATION MULTIPLICATION

Dislocation lines can be nucleated as small loops either homogeneously or heterogeneously (at grain boundaries, particle-matrix interfaces, etc.). Also, they can grow in length. Suppose they were to grow in length while lying only on their original glide planes. Then, a steady-state could occur in which dislocations were created at nucleation source and moved until they exited (disappeared) at the free surface of a crystal. The crystal would simply shear into two pieces. There would be no strain-hardening, unless secondary events occurred.

Although this sometimes occurs through the operation of Frank-Read sources it is not generally observed. What does generally occur is similar, but more complex. The process is called multiple-cross-glide, and was proposed by Koehler (1952). Its importance was first demonstrated experimentally by Johnston and Gilman (1959). In addition to its existence, they showed that the process produces copious dislocation dipoles which are responsible for deformation-hardening.

Koehler attributed the cross-gliding to thermal activation, but it was found experimentally that it increases with dislocation velocity, which is inconsistent with thermal activation, so Gilman (1997) proposed that it is associated with flutter of screw dislocations caused by phonon buffeting.

The multiple-cross-glide process does not lead to steady-state dislocation multiplication. It does lead to a proportionality between the dislocation density at N, a given time and the rate of increase of dislocation density, dN/dt, that is, to first order kinetics. Thus, the dislocation density grows exponentialy with time:

$$N = N_0 e^{\alpha t} \tag{4.1}$$

where α is a constant. This is one factor that leads to deformation-softening, if the mobile dislocation density becomes greater than the demand imposed by the applied deformation rate.

As mentioned above, multiple-cross-glide leads to dislocation dipole production. and this leads to deformation-hardening. Figures 4.4 shows how this kind of dislocation multiplication occurs, and leads to dipole formation (Gilman, 1994). After they have been formed, dipoles interact with subsequent dislocations, impeding their motion and causing strain-hardening (Gilman, 1962; also Chen, Gilman and Head, 1964). This can be described in terms of a decrease in the fraction, $f = e^{-\beta \varepsilon}$ of mobile dislocations. Then, if the concentration of mobile dislocations is N_m, Equation 4.1 becomes;

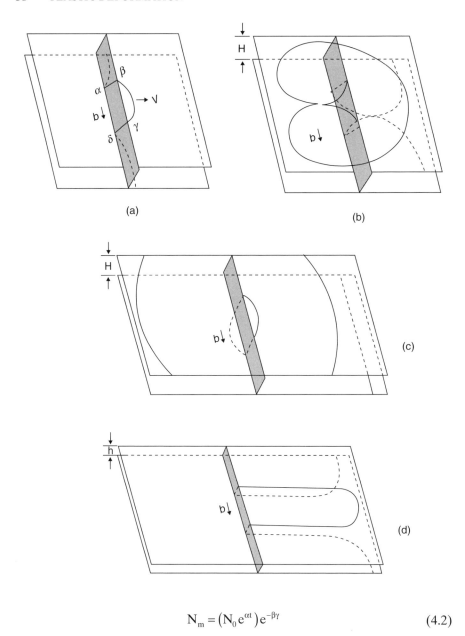

(a)

(b)

(c)

(d)

$$N_m = \left(N_0 e^{\alpha t}\right) e^{-\beta \gamma} \qquad (4.2)$$

where α and β are constants, and γ is plastic deformation. According to this equation as the total dislocation concentration grows, the mobile fraction declines. Depending on the rate of decline this results in either deformation-hardening, or deformation-softening.

Note that the dipole structure in a crystal is stabilized by the applied stress. It becomes unstable when the stress is removed. Thus, the in situ structures of

Figure 4.4   Multiplication of dislocations through multiple-cross-glide, and the production of dislocation dipoles. **a)** Because of its cylindrical symmetry, a screw dislocation is not constrained to move on its original glide plane. It can move horizontally on a plane, A, then on to a cross-plane, B (gray) and finally, back to a plane parallel to A. The jogs, αβ and δγ lie perpendicular to the Burgers vector. Therefore, they are edge-type dislocations, and are constrained to glide only on their current planes (the cross-glide plane). **b)** As the line segments, αχ, βγ, and δφ of 4.4a continue to move, βγ becomes a semi-circle and then a heart shaped configuration until the two lobes of the heart meet, and part of the pair of impinging lines annihilate. At the same time, the segments αχ and δφ swing forward on the original glide plane until they meet and partially annihilate one another. Note that the lines on the upper and lower planes must be able to pass over one another in order for the indicated events to occur. This requires that the separation of the planes, H be greater than an amount that depends on the magnitude of the applied stress. **c)** The lines of 4.4b continue to move. Segment βγ has been restored on the top plane, and there is a new segment αδ. On the bottom plane, the original line χφ is moving off to the right, and there are two segments of a new loop on the upper plane with one segment moving off to the right, and the other off to the left. Thus multiplica-tion has occurred with one line becoming three. The small loop in the center can now disappear through motion of the edge dislocations on the cross-glide plane. **d)** Showing the motion of the lines of 4.4a when the separation of the planes, h is too small for the lines to pass over one another under the given applied stress. Two edge dipoles are left behind as the segments χα and δφ move forward on the bottom plane. The segment βγ moves forward on the top plane. The applied stress tends to move the dipoles apart, but when it is removed they can move together and annihilate.

plastically deforming crystals may be quite different from their post-deformation structures. This has led to considerable confusion in the literature of the subject. It accounts, for ex-ample, for the the Bauschinger effect.

4.6   INDIVIDUAL DISLOCATION VELOCITIES (MICROSCOPIC DISTANCES)

At the microscopic length scale (circa, one micron) dislocations move intermittently. They move rapidly between obstacles, and hesitate when confronted by an obstacle. The average velocities range from less than 10^{-7} to 10^{+4} cm./sec., or eleven orders of magnitude. At the high end of this range where the applied stresses are large, the dislocations pass readily through obstacles so their velocities are limited by viscous drag. Granato (1968) has pointed out that the slope of the velocity-stress curve is then approximately equal to the drag coefficient measured at very small velocities through internal friction studies.

At high velocities (near the velocity of sound), the speed is limited by viscous drag. The conventional wisdom for many years was that the velocity is limited by inertia (the effective mass), but this is not the case (Gilman, 2000). One reason is that dislocations have zero rest mass.

An equation that describes the dependence of dislocation velocity, v on the applied shear stress, τ is:

$$v = v_0 e^{-D/\tau} \tag{4.3}$$

where v_0 is the limiting velocity, and D is a coefficient.

it has been found experimentally that when dislocations move through a region that has already been deformed, so it contains dislocations and dipoles, the velocity-stress curve is simply shifted parallel to itself to higher stresses (Gilman and Johnston, 1960). Therefore, if δ is the plastic deformation, Equation 4.3 can be modified to take this into account by changing the coefficient, D to D + Hγ. Here, H is the deformation- hardening coefficient.

For the case of LiF crystals, both the dislocation concentration and the incremental stress caused by plastic deformation are proportional to the amount of deformation. This indicates that the hardening is caused by impediments created by dislocations and dipoles to the motion of subsequent dislocations.

In metals, the incremental stress of deformation-hardening is often reported to be proportional to the square root of the dislocation density. However, In view of the mechanism of dislocation multiplication, and the subsequent deformation hardening, this is highly unlikely, so this author believes that either the data are faulty, or they are being misinterpreted.

4.7 VISCOUS DRAG

Dislocation motion is non-conservative. It occurs only when stress is applied. There is no observable "over-shoot." Thus, the applied work is converted into either heat, or additional dislocation lines (or other defects). Viscosity at the dislocation core along the glide plane determines the velocity both in the "stick" and the "slip" regimes described by Equation (4.3). This means there are at least two viscous mechanisms (Gilman, 1969): gas-like and liquid-like. In general, viscosity results from the transfer of momentum down a velocity gradient in a fluid. This slows down the faster moving fluid while speeding up the slower fluid. In a gas, particles move down the gradient (from a region of higher, to a region of lower momentum). This is the Maxwellian mode. The viscosity coefficient in this case inceases with the gas temperature.

On the other hand, when a liquid is sheared between two planes, and there is bonding with the planes, the bonds transfer momentum from the faster plane to the slower one. This is the liquid-like mode. In this case, the viscosity coefficient decreases with increasing temperature.

In pure metals at low stresses and temperatures, the gas-like mode is important, and the momentum carriers are electrons and phonons. For pure, simple metals there is essentially no shear bonding at the cores of dislocations, so the liquid-like viscosity is negligible (Kuhlmann-Wilsdorf, 1960). In contrast, considerable shear bonding occurs in covalently bonded crystals so the viscous drag is liquid-like.

4.7.1 Pure Metals

The viscosity coefficients at dislocation cores can be measured either from direct observations of dislocation motion, or from ultrasonic measurements of internal friction. Some directly measured viscosities for pure metals are given in Table 4.1. Viscosities can also be measured indirectly from internal friction studies. There is consistency between the two types of measurement, and they are all quite small, being 1–10% of the viscosities of liquid metals at their melting points. It may be concluded that hardnesses (flow stresses) of pure

TABLE 4.1   Directly Measured Damping Constants

Metal	Temperature (K)		Viscous Damping Constant	Reference
	300	4.2	$(10^{-4} P)$	
K	×		11	
		×	37	(1)
Pb		×	4.5	(2)
		×	1.5*	(2)
	×		3.4	(2)
Al	×		5.7	(3)
		×	4.8	(3)
	×		2.5	(4)
Zn	×		7.8	(7)
Cu	×		7.1	(5)
	×		1.4	(6)
	×		8.0	(10)
Fe	×		820	(1)
		×	340	(1)
		×	≈4	(8)
Sb	×		0.9	(9)

*Superconducting state

References
N. Urabe and J. Weertman, Materials Science and Engineering, **18**, 41 (1975).
V. R. Paramesaran and J. Weertman, Metall. Trans. **2**, 1233 (1971).
V. R. Paramesaran and J. Weertman, J. Appl. Phys., **43**, 2982 (1972).
J. A. Gorman, D. S. Wood, and T. Vreeland, J. Appl. Phys., **40**, 833 (1969).
W. F. Greenman, T. Vreeland, and D. S. Wood, J. Appl. Phys., **38**, 3595 (1967).
T. Suzuki, A. Ikushima, and M. Aoki, Acta Met., **12**, 1231 (1964).
D. P. Pope, T. Vreeland and D. S. Wood, J. Appl. Phys., **38**, 4011 (1967).
T. J. McKrell and J. M. Galligan, Scrip. Mat., **42**, 79 (2000).
P. P. Pal-Val, V. Ya. Platkov, and V. I. Startsev, Phy. Stat. Sol. A, **38**(1), 383 (1976).
G. A. Alers and D. O. Thompson, J. Appl. Phys., **32**, 283 (1961).

metals are not determined by intrinsic resistance of pure metals to dislocation motion. Extrinsic factors must be considered. These include: impurities (both substitutional and interstitial), grain boundaries, other dislocations, dislocation dipoles, surface coatings, domain walls, precipitates, twins, and more.

These data show clearly that that the intrinsic behavior in pure metals is visco-elastic with the velocity proportional to the applied stress (Newtonian viscosity). Although there is a large literature that speaks of a quasi-static "Peierls-Nabarro stress," this is a fiction, probably resulting from studying of insufficiently pure metals.

In impure metals, dislocation motion ocures in a "stick-slip" mode. Between impurities (or other point defects) slip occurs, that is, fast motion limited only by viscous drag. At impurities, which are usually bound internally and to the surrounding matrix by covalent bonds, dislocations get stuck. At low temperatures, they can only become freed by a quantum mechanical tunneling process driven by stress. Thus this part of the process is mechanically, not thermally, driven. The description of the tunneling rate has the form of Equation (4.3). Overall, the motion has two parts: the viscous part and the tunneling part.

4.7.2 Covalent Crystals

Since covalent bonding is localized, and forms open crystal structures (diamond, zincblende, wurtzite, and the like) dislocation mobility is very different than in pure metals. In these crystals, discrete electron-pair bonds must be disrupted in order for dislocations to move.

In these crystals, dislocation motion is divided into two regimes, above and below their Debye temperatures. Above their Debye temperatures, dislocation motion is thermally activated. The activation energies are equal to twice the band energy gaps, consistent with breaking electron-pair bonds (Figure 4.3).

When the stress (compressive) rises to a value approaching G/10 near the Debye temperature, motion of gliding dislocations tends to be replaced by the formation of phase transformation dislocations. The crystal structure then transforms to a new one of greater density. This occurs when the compressive stress (the hardness number) equals the energy band gap density (gap/molecular volume).

In tension, fracture occurs before the stress reaches G/10.

4.8 DEFORMATION-SOFTENING AND ELASTIC RELAXATION

In order to cause plastic deformation, stress must be created in a material by loading it elastically. This creates a complication by coupling elastic strains with plastic deformations, and thereby creating an interplay between elastic strain energy and the absorption of energy by plastic deformation. Situations can then exist in which elastic strain-energy drives plastic deformation without any change in the nominally applied stress. One way in which this manifests itself is as stress-deformation curves with negative slopes. Anther is as localized plastic shear bands. The reverse cannot occur, of course, because plastic deformation is non-conservative.

Plastic deformation is heterogeneous at at least two different levels of aggregation. One is the atomic level of individual dislocations. Another is the micron level where shear bands consisting of many coupled dislocations are prevalent. During a compression test, stress is applied to the specimen through a stiff elastic spring (the testing machine being the equivalent of such a spring). Then, strain-softening is manifested by a drop in the applied force (discontinuous yielding). This happens because the dislocations in the specimen are multiplying faster than is needed for the specimen's deformation to keep up with the rate of force application by the machine.

During hardness indentation, however, decreases in the applied load cannot be observed with standard indentation machines because the volume of material being plastically deformed is very small, so the amount of stored elastic energy is small. Furthermore, there is no dislocation nucleation problem at the beginning of indentation. When a sharp Vickers, or Berkovitch, indenter first touches the surface of a specimen, the area of contact is negligibly small so the stress is arbitrary large. It can homogeneously nucleate dislocations. As the initial dislocations move, they multiply. This causes strain-hardening, which increases the amount of resistance to penetration that the indenter encounters. The process continues until the work needed to cause a further increment of plastic deformation equals the work done by the applied force moving through the next increment of penetration. This confirms the idea that indentation is controlled by strain-hardening, and not by initial yielding.

In the unstrained material far from the center of an indentation, dislocations can move freely at much lower stresses than in the material near the center where the stress (and the deformation) is much larger. Thus, local plastic shear bands can form at the edges of the indenter, and do (Chaudhri, 2004). The lengths of these shear bands are often several times the size of an indentation. The leading dislocations in these bands move in virgin (undeformed) material, so they can move at lower stresses than the dislocations in the strain-hardened material near the center of an indentation.. The patterns they form are called "rosettes."

4.9   MACROSCOPIC PLASTIC DEFORMATION

Microscopic mechanisms of plastic deformation are far too complex to be described in detail. Many attempts have been made, but they have all had a variety of shortcomings. Part of the problem is that several important deformation mechanisms involve atomic interactions which interact with one another, so not only must the interactions be described by means of quantum mechanics, but also ordinary statistical mechanics cannot be applied. Therefore, a very rough statistical approach must suffice.

In general terms, as has already been mentioned, plastic deformation is a transport process analogous with electrical and thermal conductivity. These involve an entity to be transported, a carrier that does the transporting, and a rate of transport. In the case of electrical conductivity, charge is the transport entity, electrons (or holes) are the carriers, and the electron net velocities determine the rate. In the case of plastic deformation, displacement, b (cm) is the transport entity, dislocations are the carriers, N (#/cm^2), and their velocities, v (cm/sec) determine the shear deformation rate, dδ/dt. In two dimensions, the latter is given by the Orowan Equation:

$$d\delta/dt = bNv \tag{4.4}$$

In three dimensions, there may be more than one glide system, and the dislocation line need not be straight, and there may be more than one velocity, so this becomes:

$$\frac{d\delta}{dt} = b\oint(\vec{v}\cdot\vec{n})dl \tag{4.5}$$

where the line integral refers to all of the lines in the material.

Using average values, the density of mobile dislocations, N which increases with the deformation, may be written:

$$N(\delta) = (N_o + \alpha\delta)\exp(-\beta\delta) \tag{4.6}$$

where N_o = initial mobile density; α and β = measured constants. Also, the average velocity may be written:

$$v = v_0 \exp(-D + H\delta/\tau) \tag{4.7}$$

where v_o = maximum velocity, D = drag constant for N = N_o, H = deformation hardening constant, and τ = applied shear stress. D and H are measured constants. H and β have similar, but somewhat different, effects in reising the flow stress.

These equations are all that is needed to describe a creep test at constant stress, but to describe tensile (or compression) tests, the machine being used must be taken into account because the elastic stiffness of the machine plays an important role. See Gilman and Johnston (1962).

REFERENCES

M. M. Chaudhri, "Dislocations and Indentations," in *Dislocations in Solids—Volume 12*, Edited by F. R. N. Nabarro, and J. P. Hirth, Elsevier, Amsterdam, Netherlands, p. 447, (2004).

H. S. Chen, J. J. Gilman, and A. K. Head, "Dislocation Multipoles and Their Role in Strain-Hardening," Jour. Appl. Phys., **35**(8), 2502 (1964).

C. A. Coulson, *Valence*, p. 65, Clarendon Press, Oxford, UK (1952).

J. J. Gilman, "Debris Mechanism of Strain-Hardening," Jour. Appl. Phys., **33**, 2703 (1962).

J. J. Gilman, "Dislocation Motion in a Viscous Medium," Phys. Rev. Lett., **20**, 157 (1968).

J. J. Gilman, *Micromechanics of Flow in Solids*, p. 246, McGraw-Hill Book Company, New York, UA (1969).

J. J. Gilman, "Micromechanics of Shear Banding," Mech. Mater., **17**, 83 (1994).

J. J. Gilman, "Mechanism of the Koehler Dislocation Multiplication Process," Phil. Mag. A, **76**(2), 329 (1997).

J. J. Gilman, "The Limiting Speeds of Dislocations," Met and Mat. Trans. A, **31A**, 811 (2000).

J. J. Gilman and W. G. Johnston, "Behavior of Individual Dislocations in Strain-hardened LiF Crystals," Jour. Appl. Phys., **31**, 687 (1960).

J. J. Gilman and W. G. Johnston, "Dislocations in LiF Crystals," Sol. State Phys., **13**, 147, Academic Press, New York (1962).

A. V. Granato, "Internal Friction Studies of Dislocation Motion," p. 117, in *Dlslocation Dynamics*, Edited by A. R. Rosenfield, G. T. Hahn, A. L. Bement, and R. I. Jaffee, McGraw-Hill Book Company, New York (1968).

R. Hill, *The Mathematical Theory of Plasticity*, Clarendon Press, Oxford, UK (1950).

W. G. Johnston and J. J. Gilman, "Dislocation Velocities, Dislocation Densities, and Plastic Flow in Lithium Fluoride Crystals," Jour. Appl. Phys., **30**, 129 (1959).

J. S. Koehler, "The Nature of Work-Hardening." Phys. Rev., **86**, 52 (1952).

D. Kuhlmann-Wilsdorf, "Frictional Stress Acting on a Moving Dislocation in an Otherwise Perfect Crystal." Phys. Rev., **120**, 773 (1960).

R. G. Pearson, *Chemical Hardness*, Wiley-VCH, Weinheim, Germany (1997).

M. C. Shaw, "The Fundamental Basis of the Hardness Test," Chapter 1 in *The Science of Hardness Testing and Its Research Applications*, Edited by J. H. Westbook and H. Conrad, American Society for Metals, Metals Park, OH, USA.

D. Tabor, *The Hardness of Metals*, Clarendon Press, Oxford, UK (1951).

W. Yang, R. G. Parr, and T. Uytterhoeven, Phys. Chem. Minerals, **15**, 191 (1987).

5 Covalent Semiconductors

5.1 INTRODUCTION

In covalent crystals, the prototype bonds between the atoms are formed by pairs of electrons. Their spin vectors are anti-parallel, and their charge densites are concentrated along lines connecting the atom cores. The charge densities are not uniformly distributed along their connecting lines. In the homogeneous cases (C, Si, Ge, and Sn) the electrons tend to prefer the ¼ and ¾ positions in order to minimize their mutual repulsions. X-ray scattering studies have confirmed this. In heterogeneous cases (III-V, II-VI, and I-VII compounds) the distributions are assymmetric with the electrons tending to be near the anions (Spackman, 1991).

The orbitals containing the bonding electrons are hybrids formed by the addition of the wave functions of the s-, p-, d-, and f- types (the additions are subject to the normalization and orthogonalization conditions). Formation of the hybrid orbitals occurs in selected symmetric directions and causes the hybrids to extend like arms on the otherwise spherical atoms. These "arms" overlap with similar arms on other atoms. The greater the overlap, the stronger the bonds (Pauling, 1963).

The most common—and perhaps most important—hybrid orbitals are the tetrahdral ones formed by adding one s-, and three p- type orbitals. These can be arranged to form various crystal structures: diamond, zincblende, and wurtzite. Combinations of the s-, p-, and d- orbitals allow 48 possible symmetries (Kimball, 1940).

For inelastic shear in these crystals, the driving field is the shear stress, τ which exerts a force τb^2 on each atom (molecule) of the shear (glide) plane where b is the Burgers displacement (the shortest translation displacement). Covalent bonds lie perpendicular to this plane. Since they consist of pairs of electrons, breaking one requires that two electrons be promoted from the valence (bonding) to the conduction anti-bonding band. For each promotion, energy equal to the minimum band-gap energy, E_g is needed.

Figure 5.1 shows a schematic elevation through a kink on a screw dislocation in the diamond crystal structure. The black circles lie in the plane of the figure. The white ones lie in a plane in front of the figure, and the gray ones in a plane behind the figure. The straight lines represent electron pair bonds

Chemistry and Physics of Mechanical Hardness, by John J. Gilman
Copyright © 2009 John Wiley & Sons, Inc.

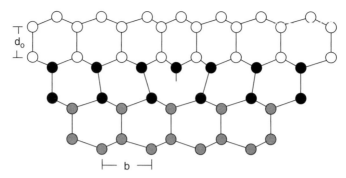

Figure 5.1   Schematic elevation view of the center of a kink on a screw disocation in the diamond crystal structure. D_o is the bond length, b is the Burgers displacement. The black circles are in the central plane of the figure. The white circles lie in a plane slightly in front of the central plane, while the gray circles lie in a plane slightly behind the central plane.

of length, d_o. One of these is "broken" at the center of the kink (a corresponding kink of the opposite sign lies elsewhere along the screw dislocation. The (111) plane lies perpendicular to the plane of the figure, and parallel to the horizontal "chains" of atoms. Movement of the kink by one Burgers displacement to the right requires that the bond to the right of the center be broken, and half of it to be combined with the initial central half-bond (Gilman, 1993). Note that all of this is highly schematic.

The shear work done for one atomic (molecular) displacement, b is the applied force times the displacement, or τb^3. This work must equal the promotion energy $2E_g$. Therefore, letting b^3 equal the molecular volume, V_m, the required shear stress is approximately $2E_g/V_m$. The parameter $[E_g/V_m]$ is called the "bond modulus." It has the dimensions of stress (energy per unit volume). The numerator is a measure of the resistance of a crystal to kink movement, while the denominator is proportional to the work done by the applied stress when a kink moves one unit distance. Overall, the bond modulus is a measure of the shear strengths of covalent bonds.

Since indentation hardness is determined by plastic deformation which is determined in turn by dislocation kink mobility, hardness is expected to be proportional to the bond modulus. Figure 5.2 shows that indeed it is for the Group IV elements, and the associated isoelectronic III-V compounds.

The discussion so far is for low temperatures; that is, temperatures below the Debye temperatures of each crystal type. There is little excitation of individual atoms below the Debye temperature. Above the Debye temperature, the temperature is associated with thermal activation and plays a much more important role, as will be discussed later.

In covalent compounds with less symmetric structures than the diamond structure factors such as ionicity, in addition to the bond moduli, need to be considered (e.g., in GaP). Surface effects (e.g., friction) also play a role in polar

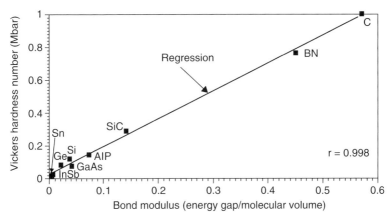

Figure 5.2 Correlation of the hardnesses of the Group IV elements, and the associated isoelectronic III-V compounds, with their bond moduli. Room temperature data. For the elements, the "molecular volumes" refer to the diatoms: C-C, Si-Si, Ge-Ge, and Sn-Sn.

crystals such as those with the zincblende structure where the α-111 surfaces on one side of a crystal are different from the β-111 surfaces on the opposite side, and there is piezoelectric activity. However, the bond modulus always seems to be the major factor in resisting inelastic shear deformation. It is a convenient parameter but is not unique because it is proportional to various other quantities associated with cohesion, for example, polarizability, plasma frequency, atomic vibrational frequencies, and elastic shear moduli.

The next two figures show that crystal structure type and ionicity also play a role in determining dislocation mobility, and therefore hardness. First, if data for the III-N compounds are plotted on Figure 5.2 they do not fall on the regression line. The reason is that they have hexagonal rather than cubic crystal structures. However, when plotted by themselves as in Figure 5.3 their hardnesses are proportional to their bond moduli.

Second, although the cubic III-V compounds lie near the regression line in Figure 5.2 they form distinct groups if plotted separately as a function of ionicity as in Figure 5.4. This figure shows that the isoelectronic III-V compounds have the same ionicities (light dotted line). Thus ionicity is one cause of the scatter in Figure 5.2.

Data for some II-VI compounds (chalcogenides) are shown in Figure 5.5.

5.2   OCTAHEDRAL SHEAR STIFFNESS

The glide planes on which dislocations move in the diamond and zincblende crystals are the octahedral (111) planes. The covalent bonds lie perpendicular to these planes. Therefore, the elastic shear stiffnesses of the covalent bonds

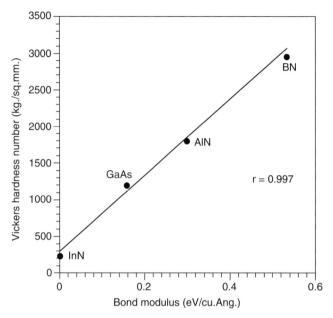

Figure 5.3   Hardnesses of the III-N nitrides vs. their bond moduli.

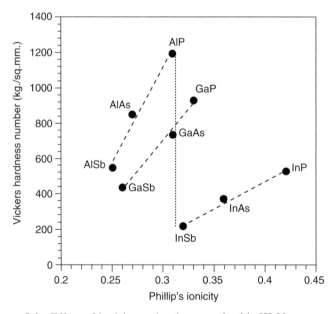

Figure 5.4   Effect of ionicity on hardnesses of cubic III-V compounds.

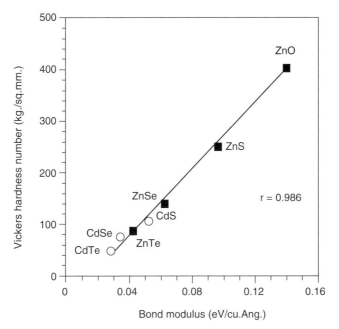

Figure 5.5   Hardnesses of some II-VI compounds (chalcogenides) versus their bond moduli.

are given by the elastic shear constants on the (111) planes. The elastic constants do not necessarily measure the inelastic bond strengths, but they play an important role in determining them. They are given by a combination of the elastic constants: C_{11}, C_{12}, and C_{44} of the cubic symmetry class:

$$C_{oct} = [3C_{44}(C_{11} - C_{12})]/[3C_{44} + (C_{11} - C_{12})] \qquad (5.1)$$

A plot of them (Figure 5.6) shows that they are proportional to the bond moduli. Thus the bond moduli are fundamental physical parameters which measure shear stiffness, and *vice versa*. Also, it may be concluded that hardness (and dislocation mobility) depends on the octahedral shear stiffnesses of this class of crystals (see also Gilman, 1973).

5.3   CHEMICAL BONDS AND DISLOCATION MOBILITY

Crystal dislocations were invented (*circa. 1930*) by Orowan, Prandtl, and Taylor to explain why pure metal crystals are soft compared with homogeneous shear strengths calculated from atomic theory. They do this very well. However, roughly 15 years later (circa 1945) it was found that pure semiconductor crystals (e.g., Ge and Si) have hardnesses at room temperature comparable with calculated homogeneous shear strengths. Furthermore, it was known

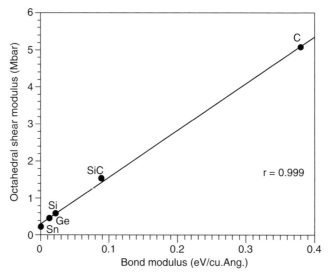

Figure 5.6 Correlation of octahedral shear stiffnesses with bond moduli for Group IV crystals. The octahedral stiffnesses measure the elastic shear resistances of the covalent bonds across the (111) planes.

that pure metal-metalloid crystals (e.g., TiC and WC) are very hard, although they conduct electricity like metals. How can these differences be explained?

The differences just outlined cannot be explained by means of a classical mechanical model. However, they can be explained by considering chemical bonding. In particular, hardness depends in covalent crystals on the fact that the valence (bonding) electrons are highly localized as shown by electron-diffraction studies which can provide maps of electron densities (bonds).

The first attempt at forming a model of dislocation mobility was that of Orowan (1965) and (Peierls, 1940). Figure 5.6 illustrates the geometry of this model. It shows two halves of a crystal: T = top, and B = bottom. The bottom plane, a of T is joined to the top plane, b of B across a glide plane. Thus the presence of B produces displacements in T (horizontal displacements only; vertical displacements are assumed to be zero), and vice versa. Start with just T. It may be considered to be an elastic body with surface tractions along a. Equal, but opposite in sign, tractions are applied along b. Therefore, the model becomes a problem in elasticity theory with the singularity at c removed by judiciously choosing the interaction potential between a and b. Since it is essentially an elastic model, and therefore a continuum model, this model cannot connect with reality for several reasons. First, because the scale of the core of a dislocation is atomic, and its is discrete rather than continuous as required by elasticity theory. The core dimensions are of the order of the wavelength of an atomic electron. Therefore, an accurate description requires quantum mechanics. Even within the context of elasticity theory. the theory is not accurate as Peierls himself discussed, as well as Orowan. For example, an

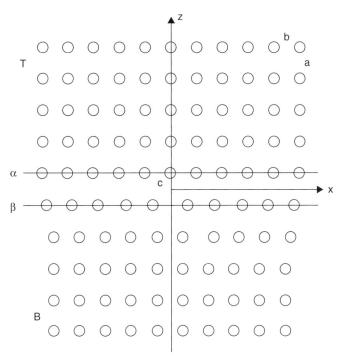

Figure 5.7   Schematic edge dislocation after Peierls. Top part of crystal, T and bottom part B, are joined between planes α and β across a glide plane with an extra half-plane of atoms ending at c. The displacement along the glide plane is b, and the glide plane spacing is a.

elastic modulus cannot be used to describe part of the interaction across the glide plane because the strains are too large for linear elasticity near the center.

A second major difficulty with the Peierls model is that it is elastic and therefore conservative (of energy). However, dislocation motion is non-conservative. As dislocations move they dissipate energy. It has been known for centuries that plastic deformation dissipates plastic work, and more recently observations of individual dislocations has shown that they move in a viscous (dissipative) fashion.

A dislocation can move no faster than its core (the region within one to two atoms of position c in Figure 5.7) so the mobility is determined by whatever barrier is presented to the core. Since the core is very localized, so must be the barrier if it is to have a substantial effect. This is why *local* covalent bonding leads to low mobility while the *non-local* bonding in metals gives high mobility.

A critical point concerns symmetry. In Figure 5.7, the atoms distant from the center have two 2-fold axes of symmetry, but the atom at the center (designated c) has quasi-5-fold symmetry. Therefore, the symmetries of the

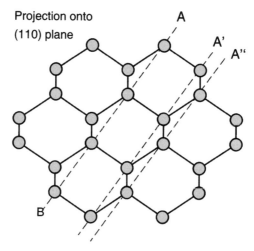

Figure 5.8   Projection of the diamond structure so the (111) glide planes (AB) are perpendicular to the plane of the figure. Then the covalent bonds connecting the atoms in planes (AB) and (A′) are perpendicular to the (111) planes. The glide plane spacing, a of the figure corresponds to the bond length AA′. The Burgers displacement, b corresponds to the atomic spacing along A or A′.

wavefunctions of the atomic electrons near the center do not match. It is a principle of quantum mechanics that wavefunctions whose symmetries do not match do not bond (Coulson, 1952). Therefore a rigorous calculation of the energy of a dislocation core is very difficult. Moreover, changes in the energy with the core's position relative to a crystal's structure are very small. Therefore, an adequately accurate calculation of the energy is exceedingly difficult. A better approach is to apply general knowledge of chemical bonding.

Dislocation motion in covalent crystals is thermally activated at temperatures above the Einstein (Debye) temperature. The activation energies are well-defined, and the velocities are approximately proportional to the applied stresses (Sumino, 1989). These facts indicate that the rate determining process is localized to atomic dimensions. Dislocation lines do not move concertedly. Instead, sharp kinks form along their lengths, and as these kinks move so do the lines. The kinks are localized at individual chemical bonds that cross the glide plane (Figure 5.8).

In the literature, it is commonly postulated that glide occurs between the planes A′ and A″ in Figure 5.8, but this seems most unlikely because the bond density is three times as large there compared with the region between A and A′.

In order for a dislocation to move along the glide region of Figure 5.8, the bonding configuration must become highly distorted and the local symmetry must change (Gilman, 1993). This can best be seen by observing a plan view of of the glide region of the diamond structure (Figure 5.9). In this figure, the larger open circles designate atoms in the plane of the paper, while the smaller

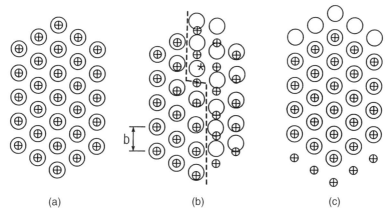

(a)   (b)   (c)

Figure 5.9 Plan view of the (111) plane of the diamond structure. A—Normal structure with open circles in the plane of the paper, and crossed circles in the plane above.
Each pair is connected by a covalent bond. B—Partial shear of the upper plane over
the lower one on the right-hand side; creating a screw dislocation line with a kink in it
(dashed line). C—Upper plane sheared down-ward by the displacement, b.

circles with crosses designate atoms lying in the plane of the crystal structure
that lies above the plane of the paper. At A, the two planes are undisturbed.
At B, the top atoms in the upper right hand quadrant have sheared downward
with the displacement b. This has created a screw dislocation with a kink in it
that is indicated by the dashed line. The kink is centered on the mid-glide
position where the displacement is b/2.

5.4 BEHAVIOR OF KINKS

A kink on a screw dislocation is a short segment of edge dislocation similar
to the configuration of Figure 5.7. By symmetry, a kink on an edge dislocation
is a short segment of screw dislocation. In both cases, the mid-glide position
contains a maximum amount of distortion of the chemical bonding. This is
indicated by the change in symmetry at the glide plane. In the normal structure,
away from the kink, each circle with a cross has six nearest neighbors in the
upper plane, one open circle below it, and six next-nearest neighbors in the
plane below it. On the other hand, at the center of the kink (indicated by
an *) the atom in the lower plane has no corresponding nearest neighbor in
the upper plane. Thus the atom in the lower plane has a valence electron with
no partner (a "broken" covalent bond). As the kink moves downward, the
broken bond reforms.

The motion of a single kink is analogous with an embedded chemical reaction of the simple exchange type (Gilman, 1993). A pair of atoms, above and
below the glide plane exchange partners when a kink moves a unit b amount.

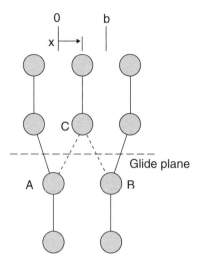

Figure 5.10 Schematic dislocation core. Arrangement at kink on screw dislocation line.

This is analogous with the classic chemical reaction in which hydrogen and deuterium exchange during a collision:

$$H_2 + D \Rightarrow HDH \Rightarrow H + DH \tag{5.2}$$

This approach is useful because it allows quantitative analysis via Walsh correlation diagrams to be made without extensive calculations. Figure 5.11 may clarify the approach. Initially the extra half-plane is at $x = 0$ and atom C is covalently bonded to atom A. When the half plane moves to the mid-glide position, $x = b/2$, the activation complex, ACB forms (Figure 5.10). Finally, when the half-plane moves to $x = b$, the pair CB forms a new covalent bond. Symbolically:

$$AC + B \Rightarrow ACB \Rightarrow CB + A \tag{5.3}$$

Here the energies of the reactants AC and B are linked (correlated) with the products, CB and A across the reaction space (reaction coordinate). When $x = b/2$, the three atoms form a transition complex, ACB.

Chemical bonds are defined by their "frontier orbitals." That is, by the highest molecular orbital that is occupied by electrons (HOMO), and the lowest unoccupied molecular orbital (LUMO). These are analogous with the top of the valence band and the bottom of the conduction band in electron band theory. However, since kinks are localized and non-periodic, band theory is not appropriate for this discussion.

The behavior of covalent semiconductors is quite different below and above the Debye temperature of a crystal. This was first shown by Trefilov and his colleagues (Gridneva, Mil'man and Trefilov, 1972). Figure 5.12 illustrates the

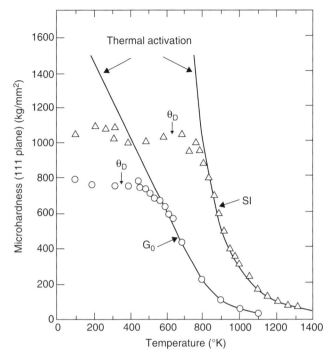

Figure 5.11   Hardness vs. temperature for Ge and Si (after Gridneva, Mil'man, and Trefilov, 1972). Showing the two regimes above and below the Debye temperatures.

behavior. It shows, for two representative cases (Ge and Si) that their hardness is independent of temperature at low temperatures, but strongly dependent on temperature at high temperatures.

A good indication of the importance of chemical bond strengths in determining hardness is the correlation between the heats of formation of compounds and their hardnesses. An example for III-V compounds is shown in Figure 5.13. The heat of formation density is equivalent to the bond modulus. This provides further evidence of the importance of chemical bond strength in determining hardness.

5.5   EFFECT OF POLARITY

Crystals whose structures are not centrosymmetric are polar because their centers of positive charge are displaced slightly from their centers of negative charge. Examples are crystals with the wurtzite structure which have polar axes along their $\langle 0001 \rangle$ directions. Also, crystals with the zincblende structure are polar in their $\langle 111 \rangle$ directions.

The surfaces lying normal to the ends of the polar axes differ in their electronic structures because they have differing chemical species exposed.

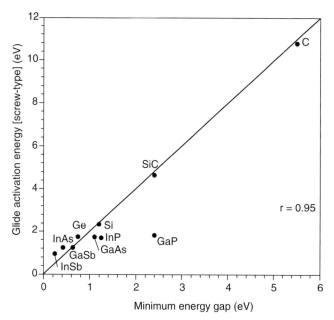

Figure 5.12 Glide activation energies from high temperature data vs. energy band gaps. Note that the data for the homopolar crystals (C, SiC, Si, and Ge) lie quite close to the correlation line, while the data for The heteropolar crystals show some scatter. The reason why GaP Is an exception is not known. Also, note that the slope of the correlation line is two.

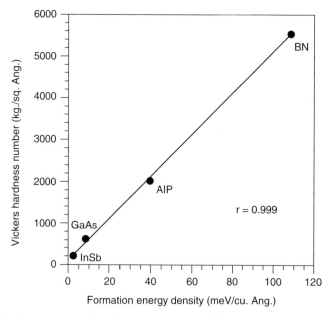

Figure 5.13 Dependence of hardness on the heat of formation density for the isoelectronic III-V compounds.

As an example, ZnO has Zn exposed at one end of the $\langle 0001 \rangle$ axis and O at the other. Therefore, there are oppositely directed molecular dipole layers at the two ends: ZnO at one end, and OZn at the other. These differing double layers lead to differing hardnesses between the two surfaces (Cline and Kahn, 1963). The difference for BeO was 1300 vs. 1100 kg/mm², or 18 percent, for indentations on the (0001) compared with the (000-1) face. In the case of ZnO the difference was 274 vs. 238 kg/mm², or 15 percent.

Similar differences have been observed for crystals with the zincblende structure, such as GaAs, on the (111) vs. the (-1-1-1) surfaces (Le Bourhis et al., 2004). However, in this case the effect is quite small; a difference of only 1.5 percent.

Structurally, the differences between the two surfaces associated with the asymmetry are localized to atomic monolayers. Therefore, the effect does not seem to involve the motion of dislocations at depths below the surfaces corresponding to the dimensions of the micro- or nano-indentations. Thus the effects may be associated with local differences such as differences in the friction coefficients at the two indenter-specimen interfaces. These coefficients may well differ because of adsorption differences. The most common adsorbed species is expected to be water molecules (Hanneman and Westbrook, 1968) although other liquids do have effects. For example, consider ZnO. On the surface terminated by Zn, chemisorption might be expected with a layer of ZnOH being formed. In contrast, on the surface terminated by O, only physisorption of H_2O might be expected. Polarization effects were also observed for GaSb, GaAs, GaP, InSb, CdTe, CdSe, ZnS, ZnSe, and ZnTe by these authors; and by Smith, Newkirk, and Kahn (1964) for BeO. The hardness changes are typically 15 to 20 percent. These polarity difference effects are not observed on dry surfaces.

Water adsorbed on surfaces is well known to substantially affect friction coefficients (Donnet et al., 1996). Other absorbates also affect hardness values.

5.6   PHOTOPLASTICITY

The effect of light on the hardness of covalent crystals was first reported by Kuczynski and Hochman (1957) for Si, Ge, and InSb. For Ge, softening of about 50 percent was observed for small loads (two grams) and the light from two 140-watt spot lamps placed 1.5 cm. from the specimen. The effect diminished with increasing depth of indentation. This latter fact suggests that the light may have been affecting the friction coefficient of the indenter/specimen interface rather than the plasticities of the specimens, probably by affecting the adsorption of water.

Later studies of the effect of light on simple compression specimens are more reliable indicators of true photoplastic effects. For example, CdS crystals were studied in compression by Garosshen, Kim, and Galligan (1990). They found that photo-plasticity required light with photon energy greater than the band gap of CdS, and the effect increased with the light intensity until a saturation level was reached. Also, the effect increases with deformation-rate, and decreases with temperature. Furthermore, the hardness is increased by the light; not decreased as in the Ge case. Finally, the effect requires time to reach its steady-state. This set of facts suggests the light creates excitons which migrate to the dislocations and inhibit their motion.

Other semiconductor crystals for which photoplasticity has been observed during hardness measurements include III-V compounds (such as GaAs— Koubaiti et al., 1997), and II-VI compounds (such as ZnS and ZnO—Klopfstein et al., 2003). However, since the effect declined in these studies with the depth of indentation, it is likely that the observations are artifacts associated with changes of the indenter/specimen friction coefficients. An extensive review of photoplastic effects in II-VI compounds is given by Osip'yan et al. (1986).

5.7 SURFACE ENVIRONMENTS

Hardness measurements of non-metallic solids are influenced by environmental factors. These have been studied extensively by Westwood (Westwood et al., 1981) and others. However, the evidence is that most, if not all, of the observed effects result from changes in the indenter/specimen friction coefficient caused by adsorption. Under ambient conditions, water vapor is commonly adsorped (Hanneman and Westbrook, 1968). In the presence of various liquids both solvents and solutes are adsorped. Since the effects are not intrinsic to the specimens, they will not be discussed further here.

5.8 EFFECT OF TEMPERATURE

For covalent crystals temperature has little effect on hardness (except for the relatively small effect of decreasing the elastic shear stiffness) until the Debye temperature is reached (Gilman, 1995). Then the hardness begins to decrease exponentially (Figure 5.14). Since the Debye temperature is related to the shear stiffness (Ledbetter, 1982) this "softening temperature" is proportional to C_{44} (Feltham and Banerjee, 1992).

5.9 DOPING EFFECTS

In the high temperature regime for covalent crystals where the hardness drops rapidly, its values are affected by impurities (dopants). Both hardening and softening occur depending on whether the dopant is is a donor, or an accepter

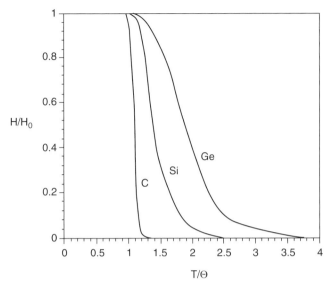

Figure 5.14 Normalized temperature dependence of the hardnesses of diamond, Si, and Ge. Note that the hardnesses divide by the low temperature hardnesses begin to decrease at the respective Debye temperatures (θ).

(Patel et al., 1974), but no effect is observed for neutral (Group IV) impurities in Ge or Si. Also, impurities that are electron-donors soften both Ge and Si at temperatures above about 450 °C; whereas accepter type impurities soften Ge, but not Si. Another important point is that small impurity concentrations have little effect. The effects begin at concentrations of about 10^{18}/cc. Since the atomic volume of Si is $20\,\text{Å}^3$, the critical ratio of impurity to Si atom is about 2×10^{-5}. Therefore, the average lineal distance between impurity atoms is about one every $270\,\text{Å}$.

REFERENCES

C. F. Cline and J. S. Kahn, "Microhardness of Single Crystals of BeO and Other Wurtzite Compounds," Jour. Electrochem. Soc., **110**, 773 (1963).

C. A. Coulson, *Valence*, Oxford University Press, Oxford, UK (1952).

C. Donnet, J. M. Martin, Th. Le Mogne, and M. Belin, "Super-low Friction of MoS_2 Coatings in Various Environments," Tribology Inter., **29**(2), 123 (1996).

P. Feltham and R. Banerjee, "Theory and Application of Microindentation in Studies of Glide and Cracking in Single Crystals of Elemental and Compound Semiconductors," Jour. Mater. Sci., **27**, 1626 (1992).

T. J. Garosshen, C. S. Kim, and J. M. Galligan, "On the Influence of Light on Dislocation Motion in Compound Semiconductors," Jour. Elect. Mater., **19**(9), 889 (1990).

J. J. Gilman, "Hardness—A Strength Microprobe," Chap. 4 in *The Science of Hardness and Its Research Applications*, Edited by J. H. Westbrook and H. Conrad, American Society for Metals, Metals Park Ohio, USA (1973).

J. J. Gilman, "Why Silicon is Hard," Science, **261**, 1436 (1993).

J. J. Gilman, "Quantized Dislocation Mobility," in *Micromechanics of Advanced Materials: A Symposium in Honor of Professor James Li's 70th Birthday*, Ed. by Chu, Liaw, Arsenault, Sadananda, Chan, Gerberich, Chau, and Kung, The Mater. Soc, Warrendale, PA, USA (1995).

I. V. Gridneva, Yu. V. Mil'man, and V. I. Trefilov, "Phase Transition in diamond–structure Crystals During Hardness Measuremnets," Phys. Stat. Sol. A, **14**(1), 171 (1972).

R. E. Hanneman and J. H. Westbrook, "Effects of Adsorption on the Indentation Deformation of Non-metallic Solids," Phil. Mag., **18**(151), 73 (1968).

G. E. Kimball, "Directed Valence", J. Chem. Phys., **8**, 188 (1940).

M. J. Klopfstein, D. A. Lucca, and G. Cantwell, "Effects of Illumination on the Response of (0001) ZnO to Nanaindentation," Phys. Stat. Sol. A, **196**, R1 (2003).

S. Koubaiti, J. J. Couderc, C. Levade, and G. Vanderschaeve, "Photoplastic Effect and Vickers Microhardness in III-V and II-VI Semiconductor Compounds," Mater. Sci. & Eng., **A234**, 865 (1997).

G. C. Kuczynski and R. F. Hochman, "Light-induced Plasticity in Semiconductors," Phys. Rev., **108**, 946 (1957).

E. Le Bourhis, G. Patrarche, L. Largeau, and J. P. Riviere, "Polarity-induced Changes in the Nanoindentation Response of GaAs," Jour. Mater. Res., **19**(1), 131 (2004).

H. Ledbetter, "Atomic Frequency and Elastic Constants," Zeit. Metallkunde, **82**(11), 820 (1982).

E. Orowan, *The Sorby Centennial Symposium on the History of Metallurgy*, Gordon and Breach, Science Publ., New York, USA (1965).

Yu. A. Osip'yan, V. F. Petrenko, A. V. Zaretskii, and R. W. Whitworth, "Properties of II-VI Semiconductors Associated with Moving Dislocations," Adv. Phys., **35**(2), 115 (1986).

J. R. Patel, "Electronic Effects on Dislocation Velocities in Heavily Doped Silicon," Phys. Rev. B, **13**, 3548 (1974).

R. E. Peierls, "The Size of a Dislocation," Proc. Phys. Soc., **52**, 34 (1940).

L. Pauling, *The Nature of the Chemical Bond—Third Edition*, Oxford University Press, Oxford, UK (1963).

D. K. Smith, H. W. Newkirk, and J. S. Kahn, "The Crystal Structure and Polarity of Beryllium Oxide," Jour. Electrochem. Soc., **111**, 78 (1964).

M. A. Spackman, "The Electron Distribution in Diamond: A Comparison between Experiment and Theory," Acta Crystallogr. A, **47**, 4200 (1991).

K. Sumino, Institute of Physics Conference, Ser. #104, p. 245, IOP, London, UK (1989).

A. R. C. Westwood, J. S. Ahearn, and J. J. Mills, "Developments in the Theory and Application of Chemomechanical Effects" (A review), Colloids and Surfaces, **2**, 1 (1981).

6   Simple Metals and Alloys

6.1 INTRINSIC BEHAVIOR

As was described above, the cohesion in simple metals depends primarily on the electron density and is nearly independent of the geometric arrangements of the atoms. This is shown by the behavior of the bulk moduli of the simple metals, the bulk modulus being a measure of the cohesive strength. Figure 6.1 shows that the bulk modulus depends primarily on the density of valence electrons. Therefore, although the energy of a dislocation line is high in pure simple metals (several eV), the energy is nearly independent of the position of the dislocation relative to the crystal structure. Thus, the energy needed to move a dislocation is negligible. That is, there is no *intrinsic resistance* to dislocation motion (Gilman, 2007).

The lack of intrinsic resistance comes from the small amount of energy needed to excite electrons. The bonding electrons are part of a sea of nearly free particles. The least energetic electrons in this sea are at the bottom of the conduction band. The most energetic are at the Fermi level. Of the many closely spaced quantum states within the conduction band, those below the Fermi level are occupied while those above the Fermi level are unoccupied (at °K). Thus only a minute fraction of an eV is needed to excite an electron from the Fermi level to an excited state. As a result, the energy of a dislocation core increases only a minute amount when it moves from one position to an adjacent one. This contrasts with the case of covalent bonding where there is a significant difference between the energy at the top of the valence band and the energy at the bottom of the conduction band.

Early in the history of crystal dislocations, the lack of resistance to motion in pure metal-like crystals was provided by the Bragg bubble model, although it was not taken seriously. By adjusting the size of the bubbles in a raft, it was found that the elastic behavior of the raft could be made comparable with that of a selected metal such as copper (Bragg and Lomer, 1949). In such a raft, it was further found that, as expected, the force needed to form a dislocation is large. However, the force needed to move a bubble is too small to measure.

At roughly the same time, Peierls (1940), and Nabarro (1947) developed a two-dimensional model of a dislocation in a simple square crystal structure. This model indicated that a small, but finite, amount of energy is needed to

Chemistry and Physics of Mechanical Hardness, by John J. Gilman
Copyright © 2009 John Wiley & Sons, Inc.

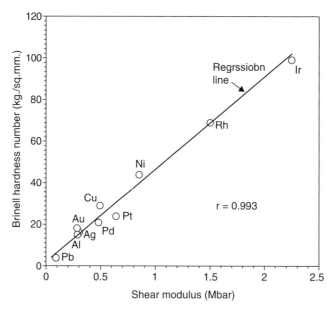

Figure 6.1   Bulk Modulus vs. Valence Electron Density (sp—bonded metals).

move a dislocation in the model crystal. Soon the conventional wisdom became that metal crystals exert a finite amount of resistance to dislocation motion. The resistance given by the Peierls-Nabarro model contains the crystal geometry in the argument of an exponential function so it is very sensitive to details of the crystal structure. Therefore, it is susceptible to "adjustments" to obtain desired outcomes. Also, no real metal crystal structures are simple "squares." And, the Peierls model is essentially a linear elastic model. Thus, the widespread acceptance of the model is surprising, especially since experimental evidence in the form of internal friction measurements made prior to its presentation cast considerable doubt on its validity. Even after direct observations of dislocation motions had been made, this model has persisted as a means to interpret a variety of experimental observations in metals. This has confused the understanding of plasticity phenomena for more than half a century.

As early as 1938, internal friction in vibrating zinc crystals was observed at strain amplitudes as small as 10^{-7}. The friction was attributed (with good cause) to dislocation motion (Read, 1938). This strongly indicated that the Peierls model could not be accepted as being quantitative.

Intrinsic resistance to dislocation motion can be measured in either of two ways: direct measurements of individual dislocation velocities (Vreeland and Jassby, 1973); or by measurements of internal friction (Granato, 1968). In both cases, for pure simple metals there is little or no static barrier to motion. As a result of viscosity there is dynamic resistance, but the viscous drag coefficient is very small (10^{-4} to 10^{-5} Poise). This is only 0.1 to 1 percent of the viscosity of water (at STP); and about 1 percent of the viscosity of liquid metals at their

TABLE 6.1   Drag Coefficients (10^{-4} Poise)

Metal	from internal friction	from velocity-stress curves
K	11	—
Al	10	2.5
Zn	—	7.8
Pb	3.7	3.4

melting points. It is consistent with the softness of high purity metals. A review of measured drag coefficients for dislocation motion has been given by Parameswaran and Arsenault (1996). Selected data for simple metals is presented in Table 6.1. The values given are for room temperature.

It is quite clear from these data that intrinsic resistance to dislocation motion in these metals does not determine their indentation hardnesses. Internal friction measurements have yielded similar results. Therefore, *extrinsic* factors need to be considered.

The commentary above refers only to pure, simple metals. It is not intended to apply to metallic compounds or to some transition metals.

6.2   EXTRINSIC SOURCES OF PLASTIC RESISTANCE

The structural materials used by engineers are not soft, but only deform plastically at large applied stresses. These result from a variety of *extrinsic* barriers to dislocation motion. Thus dislocations move freely between the barriers, but then stop until enough stress is applied to overcome the barriers.

Some of the most common extrinsic barriers are:

1. Deformation-hardening.
2. Impurity atoms (alloying).
3. Pprecipitates.
4. Grain boundaries.
5. Surface films (such as oxides).
6. Magnetic domain walls.
7. Ferroelectric domain walls.
8. Twin boundaries.

These will be briefly discussed in turn.

6.2.1   Deformation-Hardening

As screw dislocations move, since they are nearly cylindrically symmetric in simple metals, they move readily from one glide plane to another, and back

again (multiple cross-gliding). This causes them to leave behind edge dislocation dipoles. The dipoles create drag, slowing the screw's motion, and making
it rate determining (Johnston and Gilman, 1960). Also, the dipoles interfere
with the motion of subsequent dislocations causing strain-hardening (ibid,
1960). The dipoles act as traps for the subsequent dislocations, forming tripoles,
quadrupoles, and higher order multipoles (Chen, Gilman, and Head, 1964).

For one screw dislocation to move past another parallel one, where the
distance between them is h, requires a shear stress, τ:

$$\tau = Gb^2/8\pi h \tag{6.1}$$

where G is the elastic shear modulus, and b is the Burgers displacement. It is
convenient to express h as a multiple, n of b, so h = nb. Within a glide band,
the saturation dislocation density is estimated to be 10^{12} lines/cm^2 (Cottrell,
1953). Thus the lines (on average) are about 10^{-6} cm. apart, or about 40b apart.
Since cross-gliding is a random process, the h's are expected to be distributed
exponentially, so the characteristic value of n might be 40/e $\approx$ 15 (e = 2.72).
A typical value for b is 2.5×10^{-8} cm., so from Equation 6.1, $\tau_d \approx$ G/370; τ_d being
the stress for further deformation. This is approximately the observed value
of the ratio: τ_d/G.

Recall that an indentation hardness number does not measure an initial
stress for deformation in metals. It measures the stress needed for further
deformation after about 40 percent deformation has already occurred.

Figure 6.2 shows yield stress versus shear modulus data for face-centered
cubic metals at about 78 K. The yield stresses were derived from Brinell
Hardness Numbers (Gilman, 1960). The slope of the correlation line is
τ_B = G/333, in good agreement with the theoretical estimate of the previous
paragraph.

The author believes that dipoles cause deformation hardening because
this is consistent with direct observations of the behavior of dislocations in
LiF crystals (Gilman and Johnston, 1960). However, most authors associate
deformation hardening with checkerboard arrays of dislocations originally
proposed by G. I. Taylor (1934), and which leads the flow stress being proportional to the square root of the dislocation density instead of the linear proportionality expected for the dipole theory and observed for LiF crystals. The
experimental discrepancy may well derive from the relative instability of a
deformed metal crystal compared with LiF. For example, the structure in Cu
is not stable at room temperature. Since the measurements of dislocation
densities for copper are not in situ measurements, they may not be representative of the state of a metal during deformation (Livingston, 1962).

It might be argued that if dislocations are extended that cross-gliding is
inhibited. However, fast moving dislocations can decrease their kinetic energy
by reducing (or eliminating) their extension (Gilman, 2001).

Whatever mechanisms operate to cause deformation-hardening, it is
phenomenologically the most general determinant of hardness for metals.

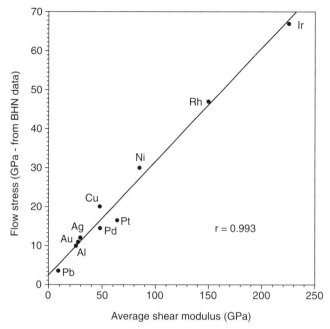

Figure 6.2 Yield stresses derived from Brinell Hardness Numbers for pure f.c.c. metals versus their shear moduli. The measurements were made at the temperature of liquid nitrogen (after Gilman, 1960).

6.2.2 Impurity Atoms (Alloying)

Most metals are used as alloys, and there are far too many of these for a comprehensive review here. A few examples will be considered, however, starting with one of the most simple of cases: Ag–Au (Gilman, 2005).

Silver and gold form a simple alloy system because they have nearly the same atomic diameters; 2.89 and 2.88 Angstroms, respectively. Both have f.c.c. crystal structures, and both come from the same column of the Periodic Table so they are isoelectronic. The two metals are mutually soluble with a heat of mixing, $\Delta U_m = -48\,\text{meV/atom}$. The molecular volume, $V_m = 8.5 \times 10^{-24}\,\text{cm}^3$, so the heat of mixing density, $\Delta U_m/V_m$ is $90.4 \times 10^8\,\text{ergs/cm}^3$.

Measurements of the yield stresses of various alloys in this system were made by Sachs and Weerts (1930). These values can be converted into hardness numbers by multiplying by three, and to shear stresses by dividing by two. The general expression for the Au concentration is $c\,(1 - c)$, where c is the concentration for each alloy. The stress needed to disrupt a Ag–Au pair is about $\Delta U_m/V_m$, and there is a maximum of these pairs when the concentration, c of Au is ½. At this maximum the hardness, H, becomes a maximum:

$$(\text{VHN})_{max} = H_{max} = 3\Delta U_m/V_m \qquad (6.2)$$

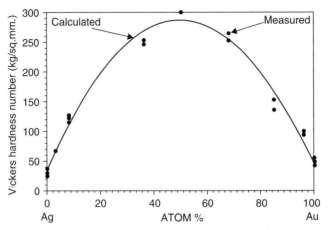

Figure 6.3 Ag–Au solid solution. Comparison of measured hardnesses and values calculated from heats of mixing. Data from Sachs and Weerts (1930).

The hardnesses of the pure metals are approximately equal, so their average is taken to be H_0. Then an expression for the hardness of an alloy may be written:

$$H = H_0 + 4(H_{max} - H_0)[c(1-c)] \tag{6.3}$$

Figure 6.3 compares the measured and calculated hardnesses for the Ag–Au alloys. The agreement is excellent. Note that the theory contains no disposable parameters. The agreement is excellent, but this may be partly fortuitous.

When normal sites in a crystal structure are replaced by impurity atoms, or vacancies, or interstitial atoms, the local electronic structure is disturbed and local electronic states are introduced. Now when a dislocation kink moves into such a site, its energy changes, not by a minute amount but by some significant amount. The resistance to further motion is best described as an increase in the local viscosity coefficient, remembering that plastic deformation is time dependent. A viscosity coefficient, η relates a rate $d\delta/dt$ with a stress, τ:

$$\tau = \eta\,(d\delta/dt) \tag{6.4}$$

so it has the dimensions force-sec/cm^2, that is, stress $\times$ time. Hence, the amount of time needed to get around, over, or through, the disturbed site increases the effective viscosity coefficient. The net result is that the presence of point defects decreases the average dislocation velocity for a given applied stress, and increases the flow stress at a given applied deformation rate.

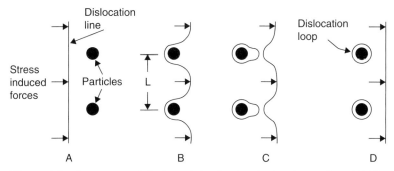

Figure 6.4   Orowan model of precipitation hardening. From Guy (1962).

6.2.3   Precipitates (Clusters, Needles, and Platelets)

When a moving dislocation line encounters a precipitate that is harder than the matrix in which it is moving, there two ways (in general) for it to get past the precipitate: Passing around it or shearing it (i.e., passing through it).

To simplify the first case, consider a pair of precipitate particles having diameters, D, and being spaced L >> D apart as in Figure 6.4. The line, A in the figure, represents an edge dislocation moving from left to right with its glide plane parallel to the plane of the figure. At A in the figure, the line approaches the precipitates. Being harder than the matrix, the precipitates are elastically stiffer than the matrix, so they repel the approaching line. Then, at B, if the applied stress is large enough, the line moves around the precipitates enveloping them. Finally, at C, the line has wrapped itself around the precipitates and has pinched off at the dotted line, leaving a loop of line around each precipitate, and begun moving onward (Orowan, 1954).

The loops around the precipitates act as stress concentrators. They exert shearing stresses in addition to the applied stress on the particles. When enough of them have accumulated, the precipitates will be plastically sheared as the loops disappear one by one. This is the basis of a theory of precipitation hardening in an aluminum-copper alloy by Fisher, Hart, and Pry (1953). The precipitate in this case is $CuAl_2$, and the precipitates cause an increment of hardening added to the hardness of the solid-solution (Al–Cu) matrix. Quantitative agreement with experimental measurements is fair.

The line can only behave in this way by being flexible. It has an energy per unit length (line tension). This energy per unit length has two parts:

1. a core energy per unit length that depends on the atomic interactions within its core. The core acts roughly like a liquid so its energy is approximately an atomic cross-sectional area, $2b^2$ times the heat of melting, ΔH_m or about 0.22 eV/atom length in aluminum.
2. The elastic strain energy of the material surrounding the core. This energy for a straight line in a large crystal is given approximately by

$U_0 = Gb^3$ where G = elastic shear modulus = 26.2 GPa. for aluminum, so the elastic part of the line energy is about 2.5 eV/atom length, or roughly ten times the core energy.

If it is assumed that the line energy, U_0, is constant (it is not, but depends on curvature of the line), and L = precipitate spacing, then there is a critical radius of curvature, r = L/2 for the dislocation line to be restrained by the pair of precipitates (Orowan, 1954). The force on the line due to the applied stress, τ is: τb per unit length, while the force resisting the bowing out of the line between the particles is: U_0/r^*, Equating the forces for the critical condition, and solving for τ gives the yield stress:

$$\tau_y = U_0/br^* = 2Gb/L \qquad (6.5)$$

so the yield stress decreases with the time of aging in the overaging regime where the particles are relatively widely spaced.

Early in the age-hardening process, the precipitates are very small, and closely spaced. Initially, they are called Guinier-Preston zones, and are only an atomic monolayer of Cu lying parallel to the (100) planes of the Al matrix (Guinier, 1994). In this regime dislocation lines are too stiff to wrap around the platelet zones. They must pass through them. Then the lines experience both positive and negative internal stresses as they move through a field of precipitates. Thus half of the time the motion of a dislocation line will be speeded up, and during the other half, it will be slowed down. Its net velocity will be decreased (Chen, Gilman, and Head, 1964). The decrease in velocity occurs because more time is lost when the local stress acts against the motion than when the local stress enhances the motion. For some values of the parameters, the incremental stress increases approximately parabolically with the aging time.

Overall, during age-hardening there are two regimes (as discussed by Orowan): an initial regime in which the yield stress rise rapidly with time (or precipitate size) before leveling off; and a second over-aging regime in which the yield stress declines to a relatively low value (as the precipitates become very large (Figure 6.5).

For extensive reviews of precipitation hardening see Brown and Ham (1971) and Ardell (1994).

6.2.4   Grain-Boundaries

Grain boundaries as barriers to dislocation motion are discussed in various text books (for example, Meyers and Chawla, 1998). The discussions either take the position that the dislocations form at the grain centers and become blocked by the dislocations of the boundaries, or that the dislocations originate at the boundaries and block one another within the grains. Here, a third view will be taken which seems more likely to this author. It is a more macroscopic

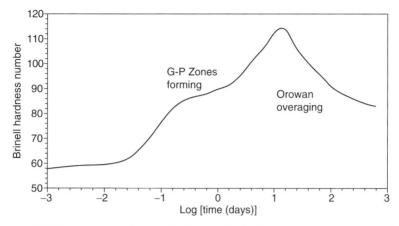

Figure 6.5 Two regimes of precipitation hardening in an aluminum-copper alloy.

viewpoint based on plastic compatibility, and is based on experiments with zinc bicrystals (Gilman, 1953).

In a polycrystalline material, a few of the grain boundaries are symmetric and therefore *plastically compatible*. Most are incompatible. What is meant hereby "compatibility" is that when one crystal within a polycrystal plastically deforms its boundaries must maintain continuity. Therefore, every small area of the deforming grain's boundary must closely match the corresponding area of its neighboring grain. Any mismatch will cause elastic strain at that location on the mutual boundary. Both the sizes of the areas and the crystal structure angles must match. This is a stringent condition. It can only be precisely satisfied at boundaries that are symmetrically disposed between two crystals.

Figure 6.6 illustrates the compatibility condition. Note that this type of compatibility is quite different from *elastic compatibility*. In Figure 6.6A, a symmetric bicrystal is shown. The dashed lines suggest the glide planes in each crystal. The two crystals of the bicrystal have thickness, t_0. The boundary is indicated by a dotted line. The half-angle between the bicrystals is θ_0. The bicrystal is shown at Figure 6.6B after some compression. Each crystal has been sheared to an angle, θ causing t_0 to become t on both sides of the grain boundary. Thus, there is compatibility at the boundary and no strain in the boundary.

Figure 6.6C contrasts with 6.6A and 6.6B by being asymmetric. The boundary no longer bisects the angle between the grains. The initial thickness remains the same at t_0, but now, when the right hand crystal shears by the angle, θ its thickness, t no longer matches the thickness of the left-hand crystal, and $t/t_0 = \cos\theta$, so a tensile strain $\varepsilon = 1 - \cos\theta$ is produced in the boundary with the corresponding stress, $\sigma = Y (1 - \cos\theta)$ where $Y = $ Young's modulus. The shear strain in the crystal is $\tan\theta$ so for relatively small strains the boundary resists further shear of the right-hand crystal and may fracture. Only a small fraction of grain boundaries in poly crystals are symmetric. Most of them

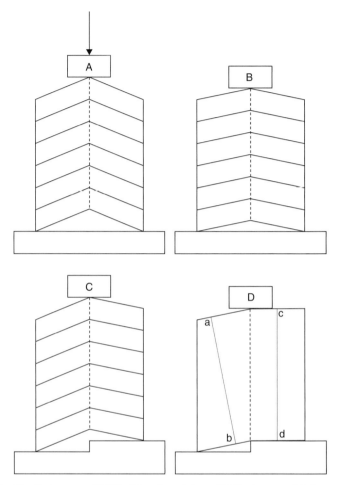

Figure 6.6   Plastic incompatibility. Showing at A and B that symmetric boundaries are plastically compatible, whereas asymmetric ones at C and D are not.

are asymmetric so this incompatibility mechanism is prevalent. Stress concentrations are needed to force deformations in glide systems other than the principal one.

Stresses can can be concentrated by various mechanisms. Perhaps the most simple of these is the one used by Zener (1946) to explain the grain size dependence of the yield stresses of polycrystals. This is the case of the shear crack which was studied by Inglis (1913). Consider a penny-shaped plane region in an elastic material of diameter, D, on which slip occurs freely and which has a radius of curvature, ρ at its edge. Then the shear stress concentration factor at its edge will be $\cong (D/\rho)^{1/2}$. The shear stress needed to cause plastic shear is given by a proportionality constant, α times the elastic shear modulus,

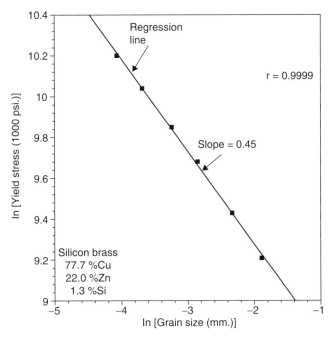

Figure 6.7 Comparison of Zener's equation with measurements of silicon-brass specimens (Data of Wilkins and Bunn, 1943).

G. Therefore, in a polycrystal, the macroscopic stress needed general plastic deformation is:

$$\sigma_y = \sigma_0 + \beta G (\rho/D)^{1/2} \quad (D \gg \rho) \quad (6.6)$$

where β is a constant, and σ_0 is the yield stress of a monocrystal of the material, and $\rho \approx b$ (an atomic diameter). Equation 6.6 is known as the Hall-Petch equation, but it was developed much earlier by Zener who found that it describes measurements of silicon-brass specimens very well (Figure 6.7).

A simple physical argument is that the shear strain energy density in a grain of a two-dimensional polycrystal with an applied stress, τ is $\tau^2/2G$, so in plane strain (remembering that shear is a 2D-process) the energy U in an average grain of diameter, D, is $U = (\tau^2/2G)(\pi D^2)$. If a freely slipping shear band is now put into the grain of length, D, with a tip radius, b, the force per unit length is $\partial U/\partial b = \pi D \tau^2/4G$ on the boundary of the band (Figure 6.8). The force resisting deformation in the surrounding material is the yield stress, $\tau_y = \alpha G$ times b. Equating the forces and solving for the applied stress, gives:

$$\tau = \beta G (b/D)^{1/2} \quad (6.7)$$

which is essentially the same as Equation (6.6).

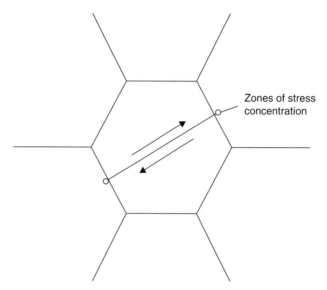

Figure 6.8   Schematic shear band in a grain surrounded by other grains. Strong concentrations of shear stress reside at each end of the glide band which has a length D (the grain diameter). The radii of curvature at the ends of the band are taken to be atomic diameters.

There is a large literature discussing the effects of grain boundaries on plastic deformation. The essential effects for "clean" boundaries have just been discussed, but there are many additional effects when the boundaries are contaminated with impurities and precipitates. All this will not be discussed further here. Books that have differing viewpoints on grain boundary effects are Baker (1983), and Meyers and Chawla (1998).

6.2.5 Surface Films (Such as Oxides)

All metal surfaces are reactive, including the noble ones. Therefore, under ambient conditions, they all have chemisorbed layers on their surfaces. These vary greatly from metal to metal in thickness, from atomic monolayers, to microns, or more. The oxide layer on gold is very thin, for example, whereas it is quite thick on copper or lead.

Oxygen is not the only surface contaminant, but it is the most prevalent. Other frequent ones are sulfur (tarnish) and phosphorus. Sometimes a combination, such as chlorate (O + Cl), forms a layer. These effects are known collectively as the "Roscoe Effect" (Metzger and Read, 1958).

Surface layers interfere with the motion of dislocations near surfaces. Among other effects, this causes local strain-hardening, creating a harder surface region which thickens with further deformation, and eventually affects an entire specimen. A specific way in which this happens is through curving

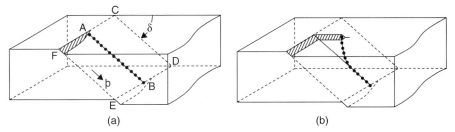

Figure 6.9 Curvature of a screw dislocation near a free surface: a. screw dislocation that has moved about halfway thru the specimen and is emerging from the surface. b. by becoming curved the screw dislocation reduces its length and hence its energy.

of screw dislocations as they approach a free surface, or interface (Gilman, 1961). Figure 6.9 illustrates this. At Figure 6.9A, a schematic screw dislocation AB (string of dots) is shown lying on a glide plane, CDEF, that makes an angle, θ, with a free surface. It has moved into the crystal from FE to AB. It can shorten, thereby lowering its energy, by curving until its end becomes normal to the surface as in Figure 6.9B. However, at the surface, the dislocation line no longer lies parallel to the Burgers vector, b. Thus, it can no longer glide in the original direction, and therefore creates drag on the motion of the disloca- tion. Also note that as a screw dislocation breaks through a surface it creates a surface step. Therefore, it interacts strongly with the free surface and with any surface film or other modification of the surface.

By increasing the surface area, screw dislocations moving at or through increase the surface area, hence the surface energy. As a result, surface active agents that affect the surface energy have an effect on near-surface screw dis- location motion (Likhtman, Rehbinder, and Karpenko, 1958). These effects are known collectively as "Rehbinder Effects." See also see other papers, for example, Westwood (1963).

Cross-gliding of screw dislocations has an important effect on the overall plastic deformations of crystals because it is the primary cause of both multiplication, and strain-hardening as discussed above.

A factor that probably plays a significant role in the surface cross-gliding mechanism is that the elastic shear energy of a screw dislocation is relaxed very near a free surface. Thus the line energy is relatively small, being deter- mined only by the core energy. This results in high flexibility of the line. The line energy is comparable with the energy of a surface step, so it is reasonable to expect that changes of surface energy will be important. This is in contrast with the case of edge dislocations where there is no step until the line emerges from a surface.

6.2.6 Magnetic Domain Walls

In their demagnetized states, ferromagnetic metals, such as iron, cobalt, and nickel, consist of randomly magnetized domains separated by well-defined domain-walls. Across these walls, the direction of magnetization changes. For example, across what are called 180° walls, the magnetization vectors in the two domains point in opposite directions making the net magnetization equal zero.

Magnetic domain walls have relatively low energies; for iron, it is 1–2 ergs/cm². (Kittel, 1956). This is miniscule compared with the free surface energy of iron, about 1800 ergs/cm². Thus these walls per se have little effect on dislocation motion, yielding, or hardness. However, magnetization also affects the dimensions of a crystal. This effect is known as magnetostriction. A result is that the elastic stiffnesses undergo small changes. In turn, these changes affect dislocation motion near the domain walls.

6.2.7   Ferroelectric Domain-Walls

In dielectric materials there can be both permanent and induced polarization domains. The walls between these domains may also act as barriers to dislocation motion. They tend to have larger energies than magnetic domain walls so they may have more effect on hardness (McColm, 1990).

6.2.8   Twin Boundaries

Twins are commonly found or formed in all types of crystals. Their boundaries are of two general types: coherent and incoherent. The coherent boundaries are usually also symmetric, so they offer little resistance to dislocation motion. However, the incoherent ones are not symmetric and may resist dislocation motion considerably.

REFERENCES

A. J. Ardell, "Intermetallics as Precipitates and Dispersoids in High Strength Alloys," Chapter 12 in *Intermetallic Compounds: vol. 2*, Edited by J. H. Westbrook and R. L. Fleischer, J. Wiley & Sons, New York, USA (1994).

T. N. Baker, Editor, *Yield, Flow, and Ftacture of Polycrystals*, Applied Science Publishers, London, UK (1983). See R. W. Armstrong, p. 1.

W. L. Bragg and W. A. Lomer, "A Dynamical Model of a Crystal Structure. II," Proc. Roy. Soc. London A, **190**, 171 (1949).

L. M. Brown and R. K. Ham, "Dislocaton-Particle Interactions," Chapter 2 in *Strenthening Methods in Crystals*, Edited by A. Kelly and R. B. Nicholson, Halsted Press Diviion of John Wiley & Sons, New York, USA (1971).

H. S. Chen, J. J. Gilman, and A. K. Head, "Dislocation Multipoles and Their Role in Strain-Hardening," J. Appl. Phys., **35**(8), 2502 (1964).

A. H. Cottrell, *Dislocations and Plastic Flow in Crystals*, p. 151, Clarendon Press, Oxford, UK (1953).

J. C. Fisher, E. W. Hart, and R. H. Pry, "The Hardening of Metal Crystals by Precipitate Particles," Acta Metal., **1**, 336 (1953).

J. J. Gilman, "Deformation of Symmetric Zinc Bicrystals," Acta Metall., **1**, 426 (1953).

J. J. Gilman, "The Plastic Resistance of Crystals," Australian Jour. Phys., **13**, 327 (1960).

J. J. Gilman, "The Mechanism of Surface Effects in Crystal Plasticity," Phil. Mag., **6**, #61, 159 (1961).

J. J. Gilman, "Mechanisms Underlying Hardness Numbers," Mater. Res. Soc. Symp. Proc., **841**, R10.11.1/T6.11.1 (2005).

J. J. Gilman, "Chemical Theory of Dislocation Mobility." Mater. Sci. and Eng., **409**, 7 (2005).

J. J. Gilman and W. G. Johnston, "Behavior of Individual Dislocations in Strain-hardened LiF Crystals," J. Appl. Phys., **31**, 687 (1960).

J. J. Gilman, "Contraction of Extended Dislocations at High Speeds," Mater. Sci. & Eng., **319A**, 84 (2001).

J. J. Gilman, "The 'Peierls Stress' for Pure Metals (Evidence That It is Negligible)," Phil. Mag., **87**, 5601 (2007).

A. V. Granato, "Internal Friction Studies of Dislocation Motion," in *Dislocation Dynamics*, Edited by A. R. Rosenfield, G. T. Hahn, A. L. Bement, and R. I. Jaffee, McGraw-Hill Book Company, New York, USA, p. 117 (1968).

A. Guinier, *X-ray Diffraction*, p. 290, Dover Publications, New York, USA (1994).

A. G. Guy, *Physical Metallurgy for Engineers*, Addison-Wesley Publishing, Reading, MA, USA, p. 271, (1962).

C. E. Inglis, *Stresses in a PlateDue to the Presence of Cracks and Sharp Corners*, Trans. Roy. Institution Naval Arch., London, **55**, 219 (1913).

W. G. Johnston and J. J. Gilman, "Dislocation Multiplication in Lithium Fluoride Crystals," J. Appl. Phys., **31**, 632 (1960).

C. Kittel, *Introduction to Solid State Physics—2nd Edition*, p. 434, J. Wiley & Sons, New York, USA (1956).

V. I. Likhtman, P. A. Rehbinder, and G. V. Karpenko, *Effect of a Surface Active Medium on the Deformation of Metals*, H. S. M. O., London, UK (1958).

J. D. Livingston, "The Density and Distibution of Dislocations in Deformed Copper Crystals." Acta Metall., **10**, 229 (1962).

I. J. McColm, *Ceramic Hardness*, Plenum Press, New York (1990).

M. Metzger and T. A. Read, Trans. AIME, **212**, 236 (1958).

M. A. Meyers and K. K. Chawla, *Mechanical Behavior of Materals*, Prentice-Hall, Inc., Upper Saddle River, New Jersey, USA (1998).

F. R. N. Nabarro, "Dislocations in a Simple Cublc Lattice," Proc. Phys. Soc., **59**, 256 (1947).

E. Orowan, in *Dislocations in Metals*, AIME, New York, USA (1954).

V. R. Parameswaran and R. J. Arsenault, "Dislocation Mobility Studies in Crystals—Four Decades in Retrospect," in *The Johannes Weertman Symposium*, Edited by R. J. Arsenault, D. Cole, T. Gross, G. Kostorz, P. K. Liaw, S. Parameswaran, and H. Sizek, The Materials Society, Warrendale, PA, USA (1996).

R. E. Peierls, "The Size of a Dislocation," Proc. Phys, Soc., **52**, 34 (1940).

T. A. Read, "The Internal Friction of Metallic Crystals." Phys. Rev, **54**, 389 (1938).

G. Sachs and J. Weerts, Zeit. f. Phys., **62**, 473 (1930).

G. I. Taylor, "Mechanism of Plastic Deformation of Crystals. I. Theoretical," Proc. Roy. Soc. (London), **145A**, 362 (1934).

T. Vreeland and K. M. Jassby, Cryst. Latt. Defects, **4**, 1 (1973).

A, R. C. Westwood, "Effects of Environment on Fracture Behavior" in *Fracture of Solids*, Edited by D. C. Drucker and J. J. Gilman, p. 553, Interscience Publishers, New York, USA (1963).

R. A. Wilkins and E. S. Bunn, *Copper and Copper Base Alloys*, McGraw-Hill Book Company, New York, USA (1943).

C. Zener, "A Theoretical Criterion for the Initiation of Slip Bands," Phys. Rev., **69**, 128 (1946).

7 Transition Metals

7.1 INTRODUCTION

The transition metals are the basis of many alloys used in structural engineering. Although their chemical behaviors form definite patterns in the Periodic Table, they can only be classified partially in their solid forms, and the interactions between them are complex. Even their crystal structures form only partial patterns. As Figure 7.1 shows, an early block of transition metals has hexagonal crystal structures. Adjacent to this is a block of bcc metals. Late in the total set is a block of fcc metals. In the middle of the total set, the elements have a variety of crystal structures. The differences in energy between these various structures are small (of order 10 meV, compared with cohesive energies of order 5 eV) so their chemical structures have subtle differences, but the differences in cohesion between the middle members of each series and the end members are large. The first long series is made complex by the ferromagnetism of Cr, Mn, Fe, Co, and Ni.

The hardnesses of metals with the same crystal structure tend to correlate with their cohesion, and one measure of the latter is their shear moduli (units = energy per unit volume). Figure 7.2 shows this for the fcc set of polycrystalline transition metals, and Figure 7.3 shows it for the set of polycrystalline bcc metals near the beginning of each of the Long Periods. It should be remembered that these hardnesses are not related to the initial mobilities of individual dislocations (yielding). They are related to the flow stress after some deformation-hardening has occurred. In this regime, dislocation motion is resisted by collisions of moving dislocations with dislocation dipoles and other multipoles (Gilman, 1960).

The Chin-Gilman parameters (H/G) are given in the figure captions. Note that the value for the bcc metals (0.02) is about five times greater than the value for the fcc metals (0.0044). Thus the bcc metals deformation harden much more rapidly than the fcc metals.

Steels and other structural transition-metal alloys are hardened by various extrinsic factors. The compositions and internal micro-structures of these materials are very complex. Therefore, simple descriptions and/or interpretations of their behaviors cannot be given, so they will not be discussed here.

Chemistry and Physics of Mechanical Hardness, by John J. Gilman
Copyright © 2009 John Wiley & Sons, Inc.

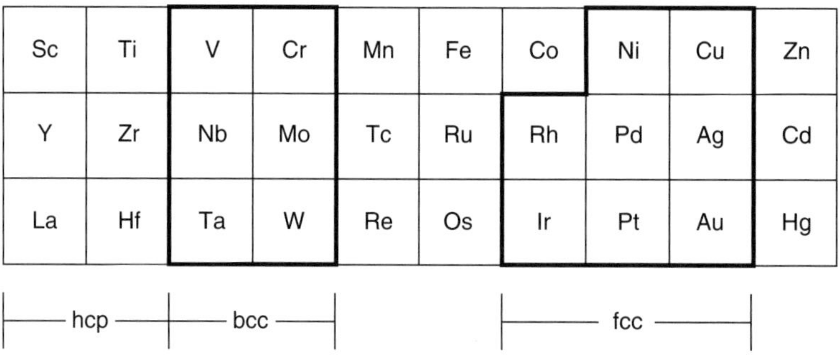

Sc	Ti	V	Cr	Mn	Fe	Co	Ni	Cu	Zn
Y	Zr	Nb	Mo	Tc	Ru	Rh	Pd	Ag	Cd
La	Hf	Ta	W	Re	Os	Ir	Pt	Au	Hg

├──── hcp ────┼──── bcc ────┤   ├──────── fcc ────────┤

Figure 7.1   Diagram showing the crystal structure groups of the transition metals.

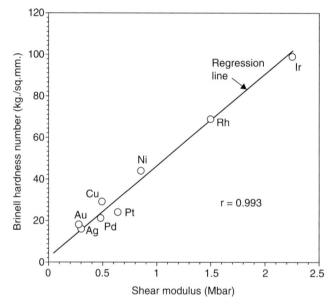

Figure 7.2   Brinell Hardness Numbers (BHN) of the fcc transition metals as a function of their average shear moduli (taken from Ledbetter, 2001). The hardness numbers are low temperature values measured at −200 °F. Note that this figure is similar to Figure 6.2 without Al and Pb. Chin-Gilman parameter = H/G = 0.0044.

Phenomenological discussion of them may be found in many other books and papers.

The most important of the extrinsic factors that affect the hardnesses of the transition metals are covalent chemical bonds scattered throughout their microstructures. These bonds are found between solute atoms and solvent atoms in alloys. Also, they lie within precipitates both internally and at precipitate interfaces with the matrix metal. In steel, for example, there are both carbon solutes and carbide precipitates. These effects are ubiquitous, but there

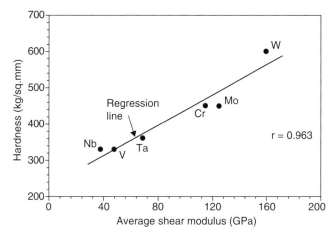

Figure 7.3 Hardness of the set of bcc transition metals as a function of their average shear moduli (taken from Ledbetter, 2001). Chin-Gilman parameter = H/G = 0.02.

is no satisfactory general theory of them, so in this author's opinion, they are best treated empirically.

7.2 RARE EARTH METALS

The rare earths have been studied extensively, but systematic trends are difficult to find, especially because they are difficult to purify. Separating them one from another is difficult because of their chemical similarity. Also, they form stable oxides so it is difficult to remove traces of oxygen from them. Much information about their hardnesses has been summarized by Scott (1978) and may be found in the handbook of Gschneidner and Eyring.

REFERENCES

J. J. Gilman, "Behavior of Individual Dislocations in Strain-Hardened LiF Crystals," Jour. Appl. Phys., **31**, 687 (1960).

H. Ledbetter and S. Kim, "Monocrystal Elastic Constants and Derived Properties of the Cubic and Hexagonal Elements," in *Handbook of Elastic Properties of Solids, Liquids and Gases.* **2**, 97 (2001).

T. E. Scott, "Elastic and Mechanical Properties," in *Handbook on the Physics and Chemistry of Rare Earths, Vol. 1, Chapter 8*, Edited by K. A. Gschneidner and L. Eyring, North-Holland Publ., New York, USA (1978–1989).

8 Intermetallic Compounds

8.1 INTRODUCTION

The hardnesses of intermetallic compounds are important not only for their own sake, but also because they are the most important hardening agents in structural alloys. They cause age-hardening. They produce wear resistance in metals for bearings. They create creep resistance at high temperatures. They make tool steels hard. They make cutlery take strong and sharp edges. And the list goes on.

Intermetallic compounds derive their great usefulness by blending metallic and covalent bonds. The former generate toughness, while the latter provide strength and hardness. In many of them dislocations move with great difficulty.

The vast array of intermetallic compounds seems limitless, although it is finite. The number of binary compounds alone is in the thousands, and the number of ternary compounds is much larger. Still larger is the number of quaternaries, quinternaries, and so on, ad infinitum. Not all of them are useful, of course, and there are many compounds having the same hardnesses as others. These overlapping cases are usually separated when other properties—such as corrosion resistance, toughness, or cost—are taken into account.

Only a few classes of compounds can be considered here. For a more comprehensive discussion the reader is referred to *Intermetallic Compounds-Vols. 1,2, and 3*, (Westbrook and Fleischer, 1995). This book considers about 2100 compounds in Vol. 1. The total number of known intermetallic compounds is at least 6000, and perhaps 11,000 or more (Villars in Westbrook and Fleischer, 1995).

A few compounds are important to society for their intrinsic hardness, although they are not as hard as borides, carbides, or nitrides. Examples are $CuAl_2$ in Duralumin for aircraft structures, Ni_3Al in Superalloys for aircraft gas turbines, and $MoSi_2$ for furnace heating elements. Given the large number of comounds, it is disappointing that so few are really useful. This is because so few are ductile.

A thorough review of the intermetallic compounds of mechanical interest has been written by Westbrook (1993). The borides, carbides, and nitrides will be considered separately in Chapter 10.

Chemistry and Physics of Mechanical Hardness, by John J. Gilman
Copyright © 2009 John Wiley & Sons, Inc.

8.2 CRYSTAL STRUCTURES

Binary intermetallic compounds crystallize in a relatively small number of structures. The simple ones are the NaCl (B1), CsCl (B2), and Ni_3Al (L1$_2$) structures, but there are also quite complex structures for transition metal binaries. An example is the "sigma phase." The prototype of this structure is the "compound," FeCr. This has a broad composition range and is the hardening agent in the XCR-steels used extensively in automotive engine valves. The sigma phase also appears in at least 45 other binary transition metal systems (Wilson and Sooner, 1973), and is a hardening phase in several of them. A few of these are CoCr, FeV, IrTa, NiV, MnMo, and ReW. The last of these has a very high melting point (nearly 3100 °C).

8.2.1 Sigma Phase

The prototype FeCr sigma phase is of particular interest because the free atoms have very nearly the same size (ratio = 1.01), but they condense into a rather intricate structure. In the pure metals, the diameter of Cr is 2.50 Å, while that of Fe is 2.48 Å. (a difference of less than one percent), and both are bcc. Therefore, the existence of the sigma phase is determined by spd-hybridization of the electron orbitals. It is sometimes called a "size-effect" phase, but this is not really descriptive.

The unit cell of the sigma phase contains 30 atoms and its symmetry class is tetrahedral. It can be built up using 2-dimensional layers having hexagonal symmetry (Frank and Kasper, 1959). These layers are Kagome nets. There are two types with hexagonal symmetry: one in which the hexagons share sides; the other in which the hexagons share corners (Figure 8.1).

The first layer in the stacking sequence of the sigma phase is type A of Figure 8.1 at level z = 0. At z = ½ is a type B layer with its "holes" aligned over the holes in the A layer. At z = 1 is another type A layer aligned with the first layer. At levels z = ¼, and z = ¾, there are atoms centered on the hole centers. These form 1D rods passing through the structure along the c-axis. The structure is quite similar to that of β-U and this is described quite well by Tucker (1951). For analysis of the packing see Frank and Kasper (1959). See also Joubert (2007).

The sigma phases are hard and brittle at below their Debye temperatures, but have some plasticity at higher temperatures. Thus there is some covalent bonding in them, and their glide planes are puckered, making it difficult for dislocations to move in them until they become partially disordered. Their structures are too complex to allow realistic hardness values to be calculated for them. Their shear moduli indicate their relative hardnesses.

The hardness of precipitated sigma phase in stainless steels seems to vary with the composition. In Type 446 plain chromium steel it is about 9 GPa (Guimaraes and Mei, 2004); whereas in Type 316 high Cr, Ni steel it is about 17 GPa (Ohmura et al., 2006). Since they are ordered phases that do

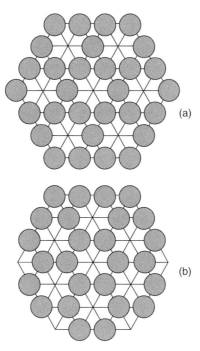

Figure 8.1   Two types of Kagome nets that form the basis of the σ—phase crystal structure. A—net with shared corners. B—net with shared edges.

not melt congruently, homogeneous sigma phase specimens are usually not available.

8.2.2   Laves Phases

When the atomic size ratio is near 1.2 some dense (i.e., close-packed) structures become possible in which tetrahedral sub-groups of one kind of atom share their vertices, sides or faces to from a network. This network contains "holes" into which the other kind of atoms are put. These are known as Laves phases. They have three kinds of symmetry: cubic (related to diamond), hexagonal (related to wurtzite), and orthorhombic (a mixture of the other two). The prototype compounds are: $MgCu_2$, $MgZn_2$, and $MgNi_2$, respectively. Only the simplest cubic one will be discussed further here. See Laves (1956) or Raynor (1949) for more details.

More than 100 intermetallic compounds have one of the Laves-type structures. Figure 8.2 illustrates the cubic $MgCu_2$ case. To form it, start with a face-centered diamond arrangement of Mg atoms (Figure 8.2a). There is a "tetroid cage" at the center of this arrangement (indicated by a star), and there are four incomplete tetroids surrounding the central one. The centers of the latter tetroids lie at the four unoccupied tetrahedral positions of the diamond structure. The $MgCu_2$ structure is generated by placing tetra-hedral clusters of

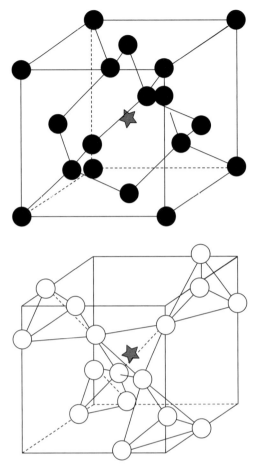

Figure 8.2 Structure (C15) of cubic Laves phases. MgCu$_2$ is the prototype. Top—Mg sub-structure with the pattern of the diamond structure. Bottom—Cu sub-structure with four tetrahedral clusters in the tetroid holes of the diamond structure. The stars indicate the centers of the patterns.

Cu atoms at each of the four empty tetrahedral positions. A corner from each of these clusters form the corners of a central tetrahedron. Thus the Cu atoms form a cluster of clusters as in Figure 8.2b.

For the Cu tetrahedra to fit into the empty spaces of the Mg pattern there must be a significant difference in the atomic diameters. In this case, the diameter ratio of the pure metals is about 3.2/2.56 = 1.25 which is just enough. Figure 8.3 is a schematic of the complete unit cell. This structure is often described in terms of layers lying normal to the ⟨111⟩ directions, but the present method is preferred by this author.

The other prototype Laves phases, MgZn$_2$ and MgNi$_2$, are formed similarly but have different symmetries. MgZn$_2$ is hexagonal and derived as described

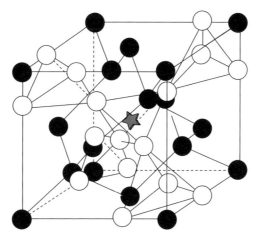

Figure 8.3 Contiuation of Figure 8.2 showing super-position of the two sub-structures to yield the C15 Laves structure. The star indicates the center.

above from the wurtzite structure. $MgNi_2$ is a mixture of the other two structures.

Since these structures are formed by filling the open spaces in the diamond and wurtzite structures, they have high atomic densities. This implies high valence electron densities and therefore considerable stability which is manifested by high melting points and elastic stiffnesses. They behave more like metal-metalloid compounds than like pure metals. That is, like covalent compounds embedded in metals.

Through X-ray scattering studies of the electron densities in $MgCu_2$, Kubota et al. (2000) found concentrations of electrons between the Cu atom pairs, but not between Mg–Cu pairs. They interpreted this as p^3d^3 covalent hybrid Cu–Cu bonds embedded in Mg metal.

An especially hard and stable Laves-type compound is cubic HfW_2. Its melting point is 2650 °C, and its hardness at room temperature is 1900 kg/mm^2 (Stone, 1977). However, it has a high mass density, so its usefulness is limited.

The Laves phases resist plastic deformation at low temperatures. However, like semiconductors they soften as the temperature increases. For $MgCu_2$, Livingston, Hall, and Koch (1988) found plastic yielding at about 0.25 GPa at 600 °C with brittleness at lower temperatures, and rapid further softening at higher temperatures. This corresponds to the expected behavior with covalent bonding. The glide system is that of fcc pure metals; (111)⟨110⟩. Glide appears to be resisted by the Cu tetrahedra.

8.2.3   Ni$_3$Al

This compound is critical to modern aircraft engines. It strengthens the alloys that are used to make the turbine blades and vanes in the highest temperature

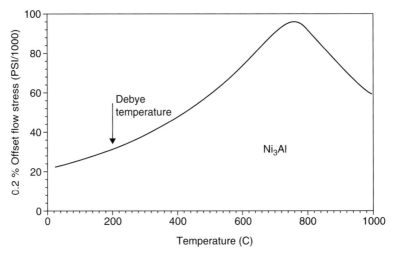

Figure 8.4   Anomalous flow stresss of nickel aluminide. Adapted from Thornton et al., 1970.

stages of the turbines. These are cast as monocrystals in order to eliminate weak grain boundaries from these engine components. The compounds, and the alloys which they strengthen, have the somewhat anomalous property of getting harder with increasing temperature in the range of interest (approx. 100–750 °C). See Figure 8.4.

The structure of Ni_3Al is the $L1_2$ (Cu_3Au) structure (Figure 8.5). It is fcc with the corners occupied by Al atoms, and the face-centers by Ni atoms. The primary glide planes are (111) and the glide directipns are ⟨110⟩. Therefore, the shears in the cores of dislocations in these crystals are broken into four parts as illustrated in Figures 8.6, 8.7, and 8.8. Each unit dislocation in the structure is split into four partial dislocations.

The arrangement of Al and Ni atoms on one side of the octahedral primary glide plane is shown in Figure 8.6. The arrangement on the other side is the same, but displaced by 0.82δ (δ = atomic diameter). The Burgers displacement vector is **b** ⟨110⟩. It is the sum of four partial vectors of the ⟨112⟩ type. That is: $\mathbf{b} = \Sigma\ \mathbf{b}_i$ (i = 1 – 4). The figure also shows two schematic glide-plane unit cells. One in the glide-plane (layer 0), and the other in the plane just above (layer 1).

In Figure 8.7 a (111) glide plane (layer 0) is shown with a unit cell from level 1 superimposed on it. Figure 8.8 shows a glide path for a layer 1 cell, as well as a unit Burgers vector. At the core of a dislocation the level 1 cell glides (shears) over the underlying plane in one of the $\mathbf{b}_i$ directions. It takes four steps for the path to become complete so there are four stacking faults and four partial directions within it.

These extended dislocations cannot move concertedly, so kinks must form on the partials and these do the moving, causing consequent movement of the

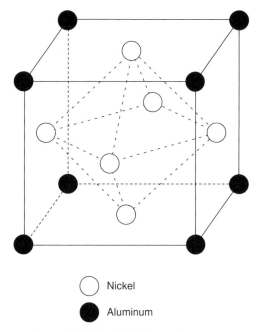

○ Nickel

● Aluminum

Figure 8.5   The Ni_3Al crystal structure.

unit dislocations. Unit kinks are not seen experimentally (at least not clearly) so the partial kinks probably repulse on another, thereby becoming distributed along the unit dislocations. Overall, then, motion of the cores in this structure is quite complex at the atomic level.

The Ni octahedra derive their stability from the interactions of s, p, and d electron orbitals to form octahedral sp^3d^2 hybrids. When these are sheared by dislocation motion this strong bonding is destroyed, and the octahedral symmetry is lost. Therefore, the overall (0 °K) energy barrier to dislocation motion is about $C_{oct}/4\pi$ where C_{oct} = octahedral shear stiffness = $[3C_{44}\ (C_{11} - C_{12})]/[4C_{44} + (C_{11} - C_{12})] = 50.8\,GPa$ (Prikhodko et al., 1998), and the barrier = 4.04 GPa. The octahedral shear stiffness is small compared with the primary stiffnesses: $C_{44} = 118\,GPa$, and $(C_{11} - C_{12})/2 = 79\,GPa$. Thus elastic as well as plastic shear is easier on this plane than on either the (100), or the (110) planes.

The coefficient, η, of the viscosity resisting dislocation motion is the shear stress at the glide plane, τ divided by the frequency of momentum transfer, v. The maximum value that τ can have is about $C_{oct}/4\pi$, and as mentioned above $v = 10^{13}/sec$ for the Al atoms, so $\eta \approx C_{oct}/4\pi v \approx 4 \times 10^{-3}$ Poise. This is comparable to the dislocation viscosity coefficients in other metallic systems. Another view of the viscosity is Andrade's theory in which:

$$\eta = (4/3)(m\ v/\delta) \qquad (8.1)$$

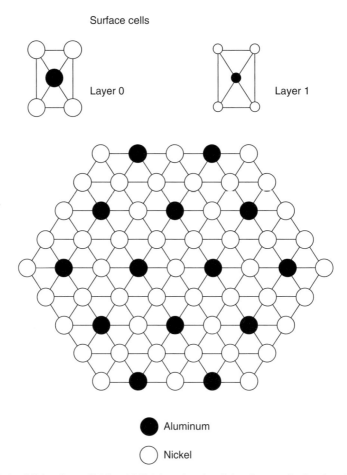

Surface cells

Layer 0

Layer 1

● Aluminum

○ Nickel

Figure 8.6   Glide plane (111) of Ni₃Al and unit glide-plane cells for levels (0—the plane below the plastic shear) and (1—the plane above the plastic shear).

with v in the numerator, where m = mass of the momentum transferring particle = average atomic mass = 71.4×10^{-24} gm; δ = glide plane spacing = 2.06 Å; and $v = 10^{13}\,\text{sec}^{-1}$. Then $\eta = 4.6$ centi-poise which is about the viscosity of the alloy at its melting point.

The observed flow stress of stoichiometric Ni₃Al at room temperature is 22 kpsi [0.156 GPa] according to Thornton, Davies, and Johnston (1970). Above room temperature the flow stress rises to a peak of about 96 kpsi. [0.676 GPa] at 1025 °K. (Figure 8.4). The increment is about 0.52 GPa.

The lattice parameter of Ni₃Al is 3.572 Å so the unit Burgers displacement is 5.052 Å, and the molecular volume is 45.58 Å³. The heat of formation is 42 kJ/mol (Rzyman et al., 1996) so $\Delta H_f = 0.49$ eV/molecule. Then $\Delta H_f/V_m = 1.7$ GPa and the flow stress might be expected to be twice this, or 3.4 GPa. However, according to Thornton et al., (1970) the 2% offset yield stress = 20 kpsi at RT = 0.14 GPa which increases to 100 kpsi. at 750 C = 0.70 GPa.

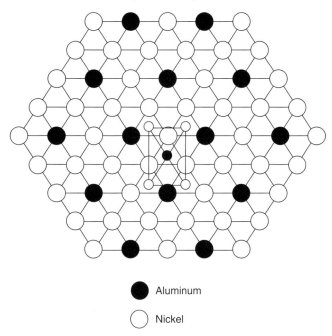

Initial position of layer 1 cell

Aluminum

Nickel

Figure 8.7 Glide plane of Ni$_3$Al with the initial position of a unit cell shown.

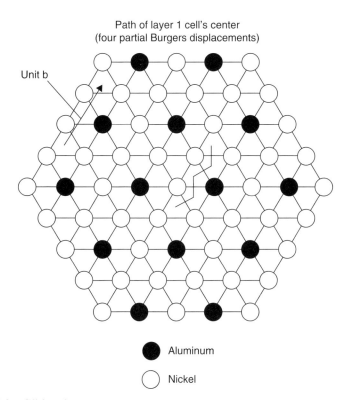

Path of layer 1 cell's center
(four partial Burgers displacements)

Unit b

Aluminum

Nickel

Figure 8.8 Glide plane of Ni$_3$Al showing paths associated with the four partial displacements that make a unit displacement.

TABLE 8.1

% Al	VHN (GPa)
24	2.16
25	2.25
26	2.04

If the bonding were covalent the expected hardness number would be about $C_{oct}/8 = 6.4$ GPa at RT, or 46X larger than the measured yield stress. It may be concluded that the bonding is essentially metallic.

Some measured values of hardness are given in Table 8.1 which shows how the hardness varies with stoichiometry (Qian and Chou, 1989). The values in the table are averages of 30 measurements for each composition. The stoichiometric value is 16X the yield stress (albeit from different authors). Since hardness numbers for metals are determined by deformation-hardening rates, the latter is very large for Ni_3Al causing the hardness numbers to be 16X the compressive yield stress instead of the 3X of pure metals.

A clue to the origin of the anomalous increase of the flow stress with temperature is given by the fact that it begins at approximately the Debye temperature. This suggests that it is associated with atomic vibrations since this is the temperature at which short wavelength phonons become excited. In this compound the vibrational frequencies of the Al atoms are quite different from those of the Ni atoms. At room temperature the Al frequencies are about 10 THz, or about twice the 5 THz frequencies of the Ni atoms. These values come from neutron diffraction studies (Stassis et al., 1981). The ratio remains about the same as the temperature rises. Therefore, the (111) glide planes become increasingly rough in terms of frequencies as the temperature rises. Notice that Equation 8.1 indicates the viscosity coefficient increases with increasing vibrational frequency because more momentum gets transferred per unit time. The vibrational Al and Ni amplitudes are approximately the same at each temperature.

Eventually the average vibration amplitudes become large enough to saturate the effect as a result of increased entropy. Then the normal decline of flow stress with temperature begins.

Another possibility is that the vibrational frequency difference increases the cross-gliding rate, and therefore the deformation-hardening rate. In this case, when the temperature becomes high enough, dislocation climb causes rapid enough recovery to cancel the deformation-hardening rate.

8.3 CALCULATED HARDNESS OF NiAl

Data are insufficient for calculation of the hardnesses of most intermetallic compounds. Also, many are too complex for realistic calculations to be made. In these cases empirical correlations with shear moduli are most likely to give useful guidance. Therefore, only an example will be treated here; namely, NiAl.

The crystal structure of NiAl is the CsCl, or (B2) structure. This is bcc cubic with Ni, or Al in the center of the unit cell and Al, or Ni at the eight corners. The lattice parameter is 2.88 Å, and this is also the Burgers displacement. The unit cell volume is 23.9 Å^3 and the heat of formation is: $\Delta H_f = -71.6$ kJ/mole. When a kink on a dislocation line moves forward one-half burgers displacement, = b/2 = 1.44 Å, the compound must dissociate locally, so ΔH_f might be the barrier to motion. To overcome this barrier, the applied stress must do an amount of work equal to the barrier energy. If τ is the applied stress, the work it does is approximately τb^3 so $\tau \approx 8.2$ GPa. Then, if the conventional ratio of hardness to yield stress is used (i.e., $2 \times 3 = 6$) the hardness should be about 50 GPa. But according to Weaver, Stevenson and Bradt (2003) it is 2.2 GPa. Therefore, it is concluded that the hardness of NiAl is not intrinsic. Rather it is determined by an extrinsic factor; namely, deformation hardening.

According to the stress-deformation curve for high purity NiAl presented by Weaver, Kaufman, and Noebe (1993) the hardening-rate for deformations of about 1–3 % is about 2.1 GPa in good agreement with the hardness given above. It is concluded that deformation hardening determines the hardness of NiAl. In other words this compound behaves plastically like a metal. The crystal structure of NiAl (CsCl) results in rapid deformation hardening because both the primary glide plane and the cross-glide plane are (110) so cross-gliding occurs readily. Although the structure (CsCl) brings ionic crystals to mind, the bonding is not ionic but is predominantly metallic.

It may be expected that many intermetallic compounds will behave like metals during plastic deformation. However, some that contain covalent bonds will behave differently. In these, the size ratio tends to be 1.2 or greater. For NiAl the size ratio is 2.86/2.49 = 1.149. This may be compared with TiC (2.89/1.54 = 1.88), or TiB$_2$ (2.89/1.72 = 1.68). The latter are clearly covalently bonded.

8.4 SUPERCONDUCTING INTERMETALLIC COMPOUNDS

Among various superconductors, compounds with the A15 (Cr$_3$Si) crystal structure have the highest critical temperatures. This crystal structure has a simple relationship with the L1$_2$ structure (Ito and Fujiwara, 1994) as illustrated in Figure 8.9. When the unit cells are aggregated, the face-centered pairs of atoms form uniform chains of transition metal atoms along three orthogonal directions. This feature may be related to the relatively stable superconductivity in compounds with this structure.

By means of hardness studies, Chin et al. (1978) investigated the type of bonding in three of these compounds. They determined the Chin-Gilman ratios (the hardness number divided by the shear stiffness). Table 8.2 lists their results. The small ratios and the moderately large hardness values indicate that

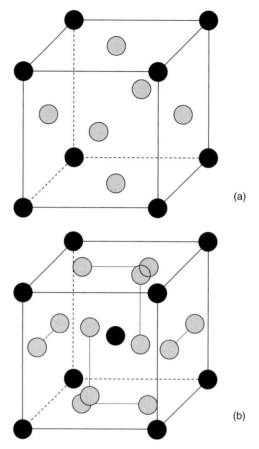

Figure 8.9 Relationship between the $L1_2$ (Ni_3Al) and the A15 (Cr_3Si) crystal structures. In both cases the cube corners are occupied by the non-transition elements (Al and Si), but the face-centers are occupied differently; by one transition metal atom in the $L1_2$ case, and by a pair of transition metal atoms in the A15 case. An additional difference is that the cube center is unoccupied in Ni_3Al, but is occupied by a Cr atom in Cr_3Si.

TABLE 8.2

Compound	C_{44} (GPa)	VHN (GPa)	Ratio
V_3Si	81.0	10.8	0.13
V_3Ge	70.0	9.7	0.14
Nb_3Sn	40.8	4.5	0.11

the bonding is covalent. Furthermore, as pointed out by Chin et al., the interatomic spacings in the A-A transition metal chains are consistent with the transition metal atoms being covalently bonded. These long and dense covalent chains embedded a metal may account for the superconductivity.

This is an example of the use of hardness measurements for interpreting other properties.

8.5   TRANSITION METAL COMPOUNDS

Numerous pairs of transition metals form compounds with various amounts of stability. Figure 8.10 shows a few of them reported by Stone (1977). These compounds form two clusters. The one at the lower left in the figure consists of pairs of metals that lie in the First Long Period (atomic numbers, 22–29), and have melting points in the range: 885–1427 °C. The second cluster consists of pairs that lie in the Third Long Period (atomic numbers, 71–77), and have melting points in the range: 2540–3160 °C.

The data of Figure 8.10 indicate two correlations. One is the high hardness is associated with high melting points. One of the factors in high melting points is high chemical stability which is also associated with high hardness. The other is that compound formation tends to occur between elements within the long periods, but not between them.

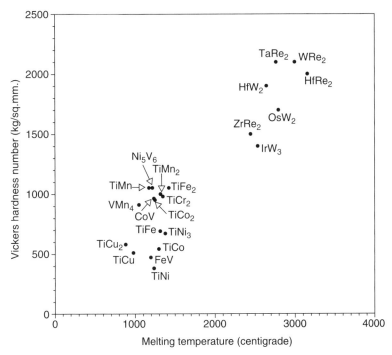

Figure 8.10   Hardnesses and melting points of some transition-metal melting points. The cluster of compounds at lower left in the figure are composed of elements from the First Long Period, while the cluster of compounds at the 1upper right are composed of elements from the Third Long Period.

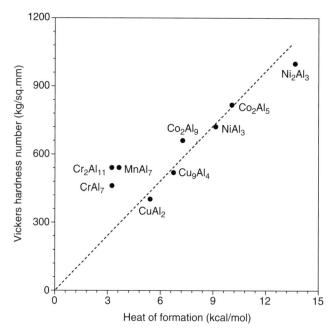

Figure 8.11 Hardnesses and heats of formation of some transition-metal aluminides. Data from E. R. Petty (1960).

In general, there are insufficient data available for quantitative estimates to be made of the hardnesses of intermetallic compounds. However, in some cases trends can be verified. Figure 8.11 illustrates one of these. It indicates that hardnesses and heats of formation tend to be related. In this case for a set of transition metal aluminides. The correlation in this case might have been improved if the heats per molecular volume couls have been plotted, but thr molecular volumes were not available. Nevertheless, the correlation is moderately good indicating that hardness and chemical bond strengths are related as in other compounds.

REFERENCES

G. Y. Chin, J. H. Wernick, T. H. Geballe, S. Mahajan, and S. Nakahara, "Hardness and Binding in A15 Superconducting Compounds," Appl. Phys, Lett., **33**, 103 (1978).

F. C. Frank and J. S. Kasper, "Complex Alloy Structures Regarded as Sphere Packings. II. Analysis and Classification of Representative Structures," Acta Cryst., **12**, 483 (1959).

A. A. Guimaraes and P. R. Mei, "Precipitation of Carbides and Sigma Phase in AISI Type Stainless Steel under Working Conditions," Jour. Mater. Process. Tech., **155–156**, 1681 (2004).

O. Ito and T. Fujiwara, "Electronic Structure Analysis of Intermetallics for Crystal Structure Changes in Nb_3Al," Modelling Simul. Mater. Sci. Eng., **2**, 363 (1994).

J.-M. Joubert, "Crystal Chemistry and Calphad Modeling of the σ Phase," *Progress in Materials Science*, Elsevier (2007).

Y. Kubota, M. Takata, M. Sakata, T. Ohba, K. Kifune, and T. Takati, "A Charge Density Study of the Intermetallic Compound $MgCu_2$ by the Maximum Entropy Method," J. Phys.: Condensed Matter, **12**, 1259 (2000).

F. Laves, "Crystal Structure and Atomic Size," p. 124 in *Theory of Alloy Phases*, Amer. Soc. Metals, Clevelnd, OH, USA (1956).

J. D. Livingston, E. L. Hall, and E. F. Koch, "Deformation and Dafects in Laves Phases." Mat. Res. Soc. Symp. Proc., **133**, *High Temperature Ordered Intermetallic Alloys III*, p. 243 (1988).

T. Ohmura, K. Tsuzaki, K. Sawada, and K. Kimura, "Inhomogeneous Nanomech-anical Properties in the Multi-phase Microstructure of Long-term Aged Type 316 Stainless Steel," Jour. Mater. Res., **21**, 1229 (2006).

E. R. Petty, "Hot Hardness and Other Properties of Some Binary Intermetallic Compounds of Aluminum," Jour. Inst. Metals, **89**, 343 (1960–61).

S. V. Prikhodko, J. D. Carnes, D. G. Issak, and A. J. Ardell, "Elastic Constants of a Ni-12.69 at. %Al Alloy from 295 to 1300 K," Scripta Mater., **38**, 67 (1998).

X. R. Qian and Y. T. Chou, "The Effect of Born ion Microhardness in Ni_3Al Polycrys-tals," MRS Symp. Proc. #133, p. 528 in *High-Temperature Ordered Inter-metallic Alloys III*, Edited by C. T. Liu et al., Mat. Res. Soc., Pittsburgh, PA, USA (1989).

G. V. Raynor, "Progress in the Theory of Alloys," p. 1 in *Progress in Metal Physics—Vol. 1*, Butterworths Scientific Publications, London, UK (1949).

K. Rzyman, Z. Moser, R. E. Watson, and M. Weinart, "Enthalpies of Formation of Ni_3Al: Experiment Versus Theory," Jour. of Phase Equilibria, **17**, 173 (1996).

C. Stassis, F. X. Kayser, C. K. Loong, and D. Arch, "Lattice Dynamics of Ni_3Al," Phys. Rev. B, **24**, 3048 (1981).

H. E. N. Stone, "Some Properties of Intertransition Metal Compounds," Jour. Mater. Sci., **12**, 1416 (1977).

P. R. Thornton, R. G. Davies, and T. L. Johnston, "The Temperature Dependence of the Flow Stress of the γ′ Phase Based on Ni_3Al," Metall. Trans, AIME, **1**, 207 (1970).

C. W. Tucker, "The Crystal Structure of the β Phase of Uranium," Acta Cryst., **4**, 425 (1951).

M. I. Weaver, M. J. Kaufman, and R. D. Noeb, "The Effects of Alloy Purity on the Mechanical Behavior of Soft Oriented NiAl Single Crystals," Scripta Met. & Mat., **29**, 1113 (1993).

M. I. Weaver, M. E. Stevenson, and R. C. Bradt, "Knoop Hardness Anisotropy and the Indentation Size Effect on the (100) of Single Crystal NiAl," Mater. Sci. Eng. A, **345**, 113 (2003).

J. H. Westbrook and R. L. Fleischer, Editors, *Intermetallic Compounds—Vol. 1*, J. Wiley & Sons, New York, USA (1995).

J. H. Westbrook, "Structural Intermetallics: Their Origins, Status and Future," in *Structural Intermetallics*, Edited by R. Darolia, et al., The Minerals, Metals & Matrials Society, Warrendale, Pa, USA (1993).

C. G. Wilson and F. J. Spooner, "A Sphere-Packing Model for the Prediction of Lattice Parameters and Order in σ Phases," Acta Cryst., **39A**, 342 (1973).

9 Ionic Crystals

9.1 ALKALI HALIDES

These crystals form a moderately large set of all combinations of the alkali metals: Li, Na, K, Rb, and Cs; and the halogens: F, Cl, Br, and I (Fr and At are usually left out of the set). Thus the set consists of $5 \times 4 = 20$ compounds.

The bonding in these compounds is largely electrostatic. It is based on the net attraction between an array of positive metal ions and an equal number of negative halogen ions. Small sets of ions are relatively unstable so this form of cohesion is dependent on long range electrostatic interactions as well as those of short range. The ions are formed by the transfer of valence electron from the alkali metal atoms to the halogen atoms so the metal ions are positive while the halogen ions are negative.

The interactions of nearest neighbors in ionic crystals are very important if the neighboring ions have the same charge signs. Therefore, for example, dislocations move in these crystals only in directions for which shear does not juxtapose ions of the same sign. These are the $\langle 110 \rangle$ directions in the NaCl crystal structure. Also, dislocations move readily on (110) planes, but not on (100) planes. This has an important bearing on hardness. Pure ionic crystals loaded on compression (or tension) deform very easily on the (110) planes because there is no interference between ions. However, for plastic indentation, the deformation is more complex, so more than one glide system is required. Glide in the additional systems cause ions of like sign to interfere. These interferences determine the hardnesses.

The hardnesses of alkali halide crystals are particularly sensitive to impurities having different electrostatic charges than the host ions because such impurities cause severe local disturbances. Thus, in crystals such as NaCl, KBr, and LiF monovalent impurities have relatively small effects, but divalent impurities have large effects, and trivalent impurities have even larger effects. Impurities may also have different polarizabilities from the host ions (the larger the ionic size, the larger the polarizability). This changes the local energy density.

Because of the sensitivity to impurities, hardness measurements within the 20 member set tend not to be systematic. One trend that is clear, however, is that hardness decreases with increasing polarizabilty (Figure 9.1).

Chemistry and Physics of Mechanical Hardness, by John J. Gilman
Copyright © 2009 John Wiley & Sons, Inc.

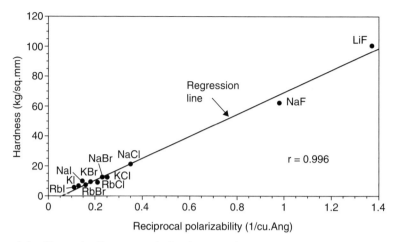

Figure 9.1 Shows the linear correlation between hardness and reciprocal polarizability for 11 alkali halides. The polarizability data are from Ruffa (1963), and the hardness data from Sirdeshmukh et al. (1995).

Another special factor in ionic crystals is that dislocation cores in them acquire net charge. As a result, plastic bending of an ionic crystal causes the top and bottom regions to become charged relative to the middle. This is easily demonstrated because such specimens preferentially attract fine insulating powders. It has been studied in some detail by Li (2000).

Figure 9.2 is schematic diagram of the crystal structure of most of the alkali halides, letting the black circles represent the positive metal ions (Li, Na, K, Rb, and Cs), and the gray circles represent the negative halide ions (F, Cl, Br, and I). The ions lie on two interpenetrating face-centered-cubic lattices. Of the 20 alkali halides, 17 have the NaCl crystal structure of Figure 9.1. The other three (CsCl, CsBr, and CsI) have the "cesium chloride" structure where the ions lie on two interpenetrating body-centered-cubic lattices (Figure 9.3). The plastic deformation on the primary glide planes for the two structures is quite different.

9.2 GLIDE IN THE NaCl STRUCTURE

The plane on which dislocations move most easily in the NaCl structure are the (110) planes. This is the exposed diagonal plane in Figure 9.2 and is designated the *primary glide* plane. The *primary glide direction* is parallel to the lines of like ions lying in the (110) planes. This deformation system keeps the ions of opposite sign adjacent during the shear of one (110) plane over another, so no electrostatic conflict arises. Thus simple compression of a NaCl type crystal along a ⟨110⟩ direction occurs readily in high purity crystals. For a more general plastic deformation (like that in plastic indentation), other glide systems must participate.

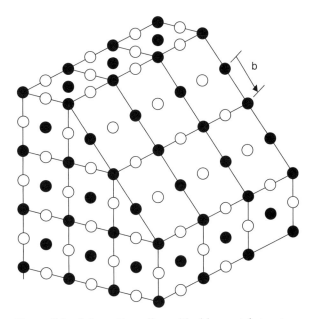

Figure 9.2   Schematic sodium chloride crystal structure.

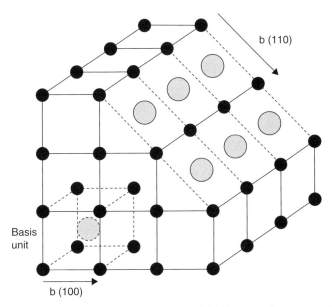

Figure 9.3   Cluster of unit cells of the cesium chloride crystal structure. This figure shows that ions of the same sign in this structure line up along the {100} directions. Thus the three rows are orthogonal to one another. Translation of a (100) plane of ions over its nearest (100) neighboring plane keeps ions of opposite sign adjacent to one another. This is also the case on the (110) planes, but the translation vector is $\sqrt{2}$ larger than for the the (100) planes.

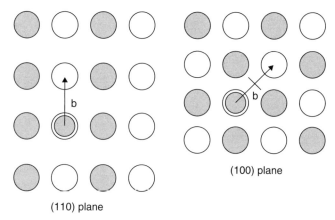

(100) plane

(110) plane

Figure 9.4 Mid-glide electrostatic faulting in the NaCl structure on the (100) planes.

For example, during indentation of the NaCl structure some compression must occur in $\langle 111 \rangle$ directions. But, if the principal stress is in a $\langle 111 \rangle$ direction, there is no shear stress on any of the primary $(110)\langle 110 \rangle$ glide systems. Therefore, a secondary system $(100)\langle 110 \rangle$ must be brought into play. The secondary system retains the primary glide direction, but creates electrostatic faults on the (100) planes at the mid-glide positions of moving dislocations (Figure 9.4). The stress needed to move a dislocation being resisted by these faults is (Gilman, 1973):

$$H = q^2 / \varepsilon b^4 = 1.2 \times 10^{-2} C_{44} \left(d/cm^2 \right) \tag{9.1}$$

here, q = electron's charge, ε = dielectric constant, b = Burgers displacement, and C_{44} = elastic shear constant.

Chin, et al. (1972) measured the hardnesses of Na and K halides (Cl, Br, and I) containing various additions of Ca^{++}, Sr^{++}, or Ba^{++}. Then they extrapolated the measurements back to zero additions to get values for the pure crystals. They found that the latter depended linearly on the Young's moduli of their crystals. Gilman (1973) found an equally good correlation with the shear stiffnesses, where $H = 1.37 \times 10^{-2} C_{44}$ (d/cm^2) in excellent agreement with Equation 9.1. A comparison of the data and the theory is given in Figure 9.5.

Since the elastic stiffness is related to the electronegativity difference density (Gilman, 2003) so is the hardness. Thus, like the covalent solids, the hardnesses of the alkali halides depends on the strength of the chemical bonding within them.

It is also worth noting that Equation 9.1 indicates a connection between C_{44}, hardness, and ε. The dielectric constant, ε depends on the polarizability, α of each alkali halide through the Clausius-Mossotti equation:

$$(\varepsilon - 1)/(\varepsilon + 1) = (4\pi/3)(\rho/MA)\alpha \tag{9.2}$$

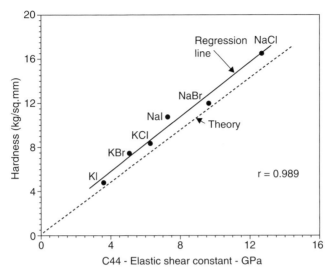

Figure 9.5 Comparison of theoretical and measured hardnesses of pure alkali halides. The solid circles are values at zero impurities extrapolated from measured values at known compositions.

9.3 ALKALI HALIDE ALLOYS

Mutual solubilities occur in many cases between pairs of alkali halides. The solubilities are often limited, but some are complete. There are 190 possible pairs for the set of 20 alkali halides. Only one particularly simple pair will be considered here, the KCl-KBr system. These two compounds in this system are completely soluble in each other. A few data points for the system have been determined by Armington, Posen, and Lipson (1973). See Figure 9.6.

A straightforward estimate of the maximum hardness increment can be made in terms of the strain associated with mixing Br and Cl ions. The fractional difference in the interionic distances in KCl vs. KBr is about five percent (Pauling, 1960). The elastic constants of the pure crystals are similar, and average values are $C_{11} = 37.5\,GPa$, $C_{12} = 6\,GPa$, and $C_{44} = 5.6\,GPa$. On the glide plane (110) the appropriate shear constant is $C^* = (C_{11} - C_{12})/2 = 15.8\,GPa$. The increment in hardness shown in Figure 9.5 is 14 GPa. This corresponds to a shear flow stress of about 2.3 GPa. which is about 17 percent of the shear modulus, or about $C^*/2\pi$.

The hardness shear modulus ratio in this case is similar to the one for metallic glasses. This suggests that the structure in the KCl-KBr solid solution is highly disordered; i.e., glassy.

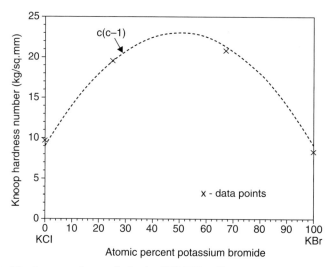

Figure 9.6 Hardnesses of crystals in the KCl-KBr alloy system. The four data points are from Armington, Posen, and Lipson (1973). The dashed curve represents the concentration function: $Ac(c-1)$ on which the data fall. c = concentration and A is a constant.

9.4 GLIDE IN CsCl STRUCTURE

Figure 9.3 shows that ions of the same sign lie parallel to the shortest transla-tion direction, $\langle 100 \rangle$ in CsCl. This is the primary glide direction, and (100) is the primary glide plane. Thus, prismatic glide can occur in any of the six $\langle 100 \rangle$ directions if a punching traction is applied to one of the (100) faces. The compressive stress from a blunt punch can push a square column through a thick crystal specimen. This is called "pencil glide."

9.5 EFFECT OF IMPURITIES

As mentioned above alkali halide crystals are strongly hardened by small additions of divalent impurities. Data are available for 12 cases, the host crystals NaCl, NaBr, KCl, and KBr with additions of Ca^{2+}, Sr^{2+}, and Ba^{2+} (Chin, et al., 1973). It was found that the hardness increases depend only on the concentrations of the additions and not on the divalent specie (Ca, Sr, or Ba). However, a dependence on the valence (1, 2, or 3) is observed. It was also found that hardness increment is proportional to the square root of the concentration, ($C^{1/2}$).

The observations given above are inconsistent with the model that Chin et al. used to interpret their measurements. This model, known as Fleischer's model, is based on the idea that the ion–ion dipoles formed between divalent

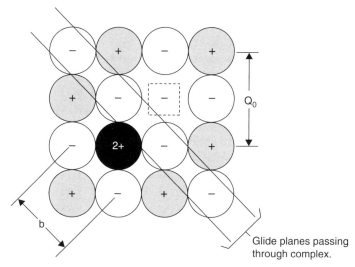

Figure 9.7   Schematic diagram of the vicinity of a divalent impurity in the NaCl crystal structure.

impurities and cation vacancies create strain fields which interact with the elastic fields of dislocations, thereby impeding the motion of the latter. This model predicts a dependence of the hardening on the host crystal's shear modulus which was not observed. Also, the differing polarizabilities of the divalent impurities should affect the local strain. This is not observed. Finally, the strain parameter, ε is used as an adjustable parameter so the theory lacks rigor.

The author proposed a model in which the charge differences are the important factor, and which has no disposable parameters. A schematic diagram of the vicinity of a divalent impurity in a monovalent ionic crystal is shown in Figure 9.7. To balance the local charges, one of the positive ions is missing. There are two possible positions of the glide planes. One passes closest to the divalent impurity. The other passes closest to the cation vacancy. The lattice parameter, a_o, and the Burgers displacement, b, are indicated on the diagram.

Before a dislocation on one of the glide planes passes through the complex, the distance between the two charge centers is $d = b = a_o/\sqrt{2}$. After it has passed by the distance is $d' = \sqrt{2}\,(b) = a_o$. Therefore, if K is the static dielectric constant, and q = electron's charge, the energy difference between the before and after states is $\Delta U = (q^2/Ka_o)(\sqrt{2}-1)$.

The work done by the applied stress, τ, is $W = (\tau/2)(b^2a_0) = (\tau/4)a_0^3$. Equating ΔU and W yields $\tau = (1.66q^2)/(Ka_0^4)$, but two glide planes pass through each complex, so the effective concentration is twice the nominal. This introduces a factor of $\sqrt{2}$. Also the compressive stress is 2τ, so the hardening coefficient is $B = (4.7q^2)/(Ka_0^4)$.

The experimental data has the form:

$$\sigma_y = A + B\sqrt{C} \tag{9.3}$$

where $A \ll B$ so it can be neglected. Using average values $K = 5.29$ and $a_0 = 6.13\,\text{Å}$. Then $B_{calc} = 11 \times 10^9\,d/cm^2$ compares with $B_{exp} = 9.2 \times 10^8\,d/cm^2$. The agreement between the two values is very good considering that no disposable parameters have been used. The good agreement between the model and the measurements model reinforces the idea that the hardness results primarily from electrostatic charge interactions, not from elastic interactions. In other words, it is an atomically localized effect.

9.6 ALKALINE EARTH FLUORIDES

The alkaline earths, Be, Mg, Ca, Sr, and Ba, and the gas Ra are divalent and form some halogen crystals under STP conditions, but they also form gaseous, liquid, and unstable halogen compounds.

The Ca, Sr, and Ba difluorides from an isomorphous set with the cubic calcium fluorite crystal structure. The other alkaline earth fluorides do not belong to this set; BeF_2 is a gas at STP, and MgF_2 is tetragonal, not cubic. Hardness numbers for these crystals are shown to be linearly proportional to their bond moduli in Figure 9.8. Similarly, their hardnesses are proportional to their inverse polarizabilities, using data from Mahbubar et al., 2002. The hardness numbers are from Klenata et al., 2005; as are the minimum (indirect) band gaps.

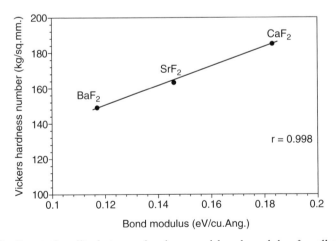

Figure 9.8 Proportionality between hardness and bond modulus for alkaline earth fluorite crystals.

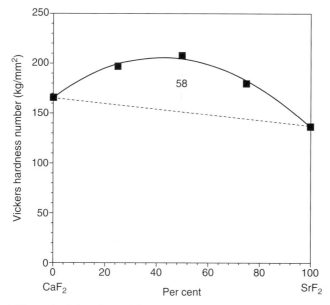

Figure 9.9   Effect of mixing Ca and Sr fluorides on hardness. The increment of hardness at 50% SrF_2 is about $58 kg/mm^2$. Data from (Chernevskaya, 1966).

The Chin-Gilman parameter for these compounds is about 0.11.

Of special interest is the behavior of solid solutions of CaF_2 and SrF_2. Figure 9.9 shows a graph of hardness vs. composition for these solutions with the peak hardness at about 50 percent SrF_2 (see also Chernevskaya, 1966). The free energy of mixing peaks at about the same concentration, but it is almost entirely entropic because the enthalpy of mixing is essentially zero. This last may be deduced because the ions are not much different in size (the lattice parameters are 5.46 and 5.86 Å for CaF_2 and SrF_2, respectively (a seven percent difference); and the shear stiffnesses of the (100) glide planes are $C_{44} = 37.5$ and 34.6 GPa, respectively (an eight percent difference); and the cations have the same charge.

The entropy of mixing (ideal solution) is:

$$S = -Nk[c \ln c + (1-c) \ln (1-c)]$$
$$= 0.693 \, Nk \qquad \text{when } c = 1/2 \qquad (9.4)$$

Thus the free energy of mixing is $-0.693 NkT$; or 1.8 meV/cation at 300 °K. The molecular volume is $a_0^3/8 \approx 20.3$ Å, so the cation volume is ≈ 6.8 Å^3. Thus the mixing free energy density is $\approx 43 kg/mm^2$ which agrees well with the $58 kg/mm^2$ increment of hardness shown in Figure 9.9.

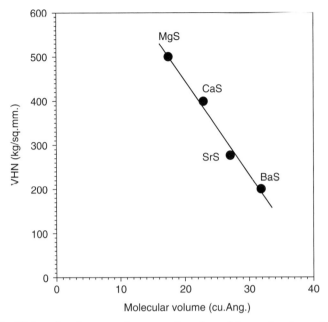

Figure 9.10 Vickers hardness numbers as a function of the molecular volumes of the alkaline earth sulfides.

9.7 ALKALINE EARTH SULFIDES

Data for ionic oxides (alkaline earth oxides) are presented in Chapter 11, although they could be presented in this chapter since they are bonded ionically. The sulfides are closely related, and are presented here.

Indentation data for the sulfides could not be found in the literature. However, Mohs scratch hardness numbers were found (Winkler, 1955). They were converted to Vickers numbers using a correlation chart. The hardnesses are shown in Figure 9.10. Since they all have the same number of valence electrons, this is the same as plotting the hardnesses versus the valence electron densities.

9.8 PHOTOMECHANICAL EFFECTS

In alkali halide crystals containing color-centers (F-centers) illumination with light of appropriate energy causes transient changes of hardness (Nadeau, 1964). This effect apparently results from changes in the effective sizes of the F-centers when they become excited.

9.9   EFFECTS OF APPLIED ELECTRIC FIELDS

As mentioned earlier in this chapter dislocations in ionic crystal may carry a net electric charge. Therefore, their motion may be influenced by applied electric fields, and may generate observable fields external to a specimen during plastic flow. These effects have been studied by Li (2000) and others.

9.10   MAGNETO-PLASTICITY

It was discovered by Al'shits et al. (1987) that static magnetic fields of order 0.5T affect the motion of dislocations in NaCl crystals. This is not an intrinsic effect but is associated with impurities and/or radiation induced localized defects. Also, magnetic field effects have been observed in semiconductor crystals such as Si (Ossipyan et al., 2004).

REFERENCES

V. I. Al'shits, E. V. Darinskaya, T. M. Perekalina, and A. A. Urusovskaya, "Motion of Dislocations in NaCl Crystals Under the Action of a Static Magnetic Field," Sov. Phys. Sol. State, **29**(2), 265 (1987).

A. F. Armington, H. Posen, and H. Lipson, "Strengthening of Halides for Infrared Windows," Jour. Electronic Mater., **2**, 127 (1973).

E. G. Chernevskaya, "The Hardness of Mixed Single Crystals of the CaF_2 Type," Sov. Jour. Opt. Tech., **33**, 346 (1966).

G. Chin, L. Van Uitert, M. Green, and G. Zydzik, "Hardness, Yield Strength and Young's Modulus in Halide Crystals," Scripta Met., **6**, 475 (1972).

G. Chin, L. G. VanUitert, M. L. Green, G. J. Zydzik, and T. Y. Kometani, "Strengthening of Alkali Halides by Divalent-Ion Additions," Jour. Amer. Cer. Soc., **56**, 369 (1973).I

P. A. Cox, *Transition Metal Oxides*, Oxford University Press, Oxford, UK (1995).

J. J. Gilman, "Hardness of Pure Alkali Halides," J. Appl. Phys., **44**, 982 (1973).

J. J. Gilman, *Electronic Basis of the Strength of Materials*, p. 137, Cambridge University Press, Cambridge, UK (2003).

R. Klenata, B. Daouda, M. Sahnoun, H. Baltache, M. Rerat, A. H. Reshak, B. Bouhafs, H. Abid, and M. Driz, "Structural, Electronic and Optical Properties of Fluorite-type Compounds," Eur. Physical Jour. B, **47**, 63 (2005).

J. C. M. Li, "Charged Dislocations and Plasto-electric Effect in Ionic Crystals," Mater. Sci. Eng., **A287**, 265 (2000).

R. M. Mahbubar, Y. Michihiro, K. Nakamura, and T. Kanashiro, "LDA Studies on Polarizabilities and Shielding Factors of Ions in Fluorite Structure Crystals," Sol. St. Ionics, **148**, 227 (2002).

J. S. Nadeau, "Two Photomechanical Effects in Alkali Halide Crystals," Jour. Appl. Phys., **35**, 669 (1964).

Yu. A. Ossipyan, R. B. Morgunov, A. A. Baskakov, A. M. Orlov, A. A. Skvortsov, E. N. Inkina, and Y. Tanimoto, "Magnetoresonant Hardening of Silicon Single Crystals," JETP Letters, **79**(3), 126 (2004).

L. Pauling, *The Nature of the Chemical Bond—3rd Edition*, p. 526, Cornell University Press, Ithaca, NY, USA (1960).

A. R. Ruffa, "Theory of the Electronic Polarizabilities of Ions in Crystals: Application to the Alkali Halide Crystals," Phys. Rev., **130**, 1412 (1963).

D. B. Sirdeshmukh, K. O. Sukhadra, K. K. Rao, and T. T. Rao, "Hardness of Crystals with NaCl Structure and the Significance of the Gilman-Chin Parameter," Cryst. Res. Technol., **30**, 861 (1995).

H. G. F. Winkler, *Struktur und Eigenshaften der Kristalle*, Springer Verlag, Berlin, Germany (1955).

10 Metal-Metalloids (Hard Metals)

10.1 INTRODUCTION

The small atoms at the center of the first row of the Periodic Table (B, C, N, O, and to a lesser extent Al, Si, and P) can fit into the interstices of aggregates of larger transition metal atoms to form boride, carbide, and nitride compounds. These compounds are both hard and moderately good electronic conductors. Therefore, they are commonly known as "hard metals" (Schwarzkopf and Kieffer, 1953).

The prototype hard metals are the compounds of six of the transition metals: Ti, Zr, and Hf, as well as V, Nb, and Ta. Their carbides all have the NaCl crystal structure, as do their nitrides except for Ta. The NaCi structure consists of close-packed planes of metal atoms stacked in the fcc pattern with the metalloids (C, N) located in the octahedral holes. The borides have the AlB_2 structure in which close-packed planes of metal atoms are stacked in the simple hexagonal pattern with all of the trigonal prismatic holes occupied by boron atoms. Thus the structures are based on the highest possible atomic packing densities consistent with the atomic sizes.

The structures of the prototype borides, carbides, and nitrides yield high values for the valence electron densities of these compounds. This accounts for their high elastic stiffnesses, and hardnesses. As a first approximation, they may be considered to be metals with extra valence electrons (from the metalloids) that increase their average valence electron densities. The evidence for this is that their bulk modili fall on the same correlation line (B versus VED) as the simple metals. This correlation line is given in Gilman (2003).

Since these compounds conduct electricity via electrons like metals, there are no gaps in their bonding energy spectra. Therefore, they do not behave like covalent compounds. However, when a kink on a dislocation line moves in them, the shearing locally disrupts the structure, thereby increasing the energy. For example, in a carbide with the rocksalt structure, the carbon atoms lie in octahedral holes. A local shear creates a stacking fault that replaces these holes with trigonal prismatic ones. The drastic change of symmetry from octahedral to tigonal prismatic decreases the local cohesion markedly because it is the orthogonal symmetry of the 2p electrons of the carbon atoms that determines the symmetries of the octahedral holes. This symmetry is also the

Chemistry and Physics of Mechanical Hardness, by John J. Gilman
Copyright © 2009 John Wiley & Sons, Inc.

131

primary reason why the prototype carbides all have the same crystal structure. Further discussion of the prototypes and other carbides is given by Cottrell (1995).

10.2   CARBIDES

The carbides with the NaCl structure may be considered to consist of alternating layers of metal atoms and layers of semiconductor atoms where the planes are octahedral ones of the cubic symmetry system. (Figure 10.1). In TiC, for example, the carbon atoms lie 3.06 Å apart which is about twice the covalent bond length of 1.54 Å, so the carbon atoms are not covalently bonded, but they may transfer some charge to the metal layers, and they do increase the valence electron density.

Primary glide occurs on the (111) planes. Shear of a carbon layer over a metal layer (or vice versa), when the core of a dislocation moves, severely disturbs the symmetry, thereby locally dissociating the compound. Therefore, the barrier to dislocation motion is the heat of formation, ΔH_f (Gilman, 1970). The shear work is the applied shear stress, τ times the molecular (bond) volume, V or τV. Thus, the shear stress is proportional to $\Delta H_f/V$, and the hardness number is expected to be proportional to the shear stress. Figure 10.2 shows that this is indeed the case for the six prototype carbides.

Another approach to relating the hardness to atomic parameters is that of Grimvall and Thiessen (1986) in which hardness is related to vibrational energies. Their theory is slightly modified here by using vibrational energy densities instead of the energies themselves. Specific heat data measure the excitation

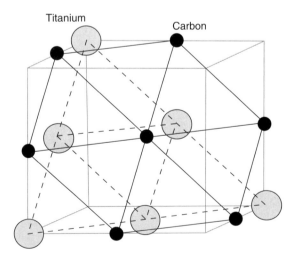

Figure 10.1   Schematic layers of Ti and C atoms in the rocksalt crystal structure of TiC.

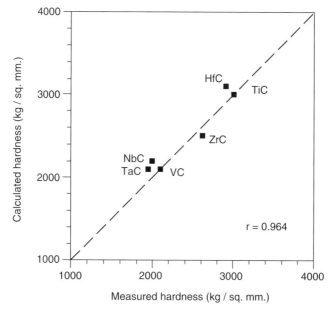

Figure 10.2   Hardnesses of prototype carbides calculated from the heats of formation vs. measured hardnesses.

of atomic vibrational modes. Since these data are known for many materials, this is a convenience of the method. There are two types of vibrational modes: longitudinal and transverse. In this case, the transverse modes are most important because they are related to the shear moduli, and these are the modes that dominate the specific heat (Ledbetter, 1991). Being related to the average shear constants, they are the ones related to indentation hardness.

To simplify the analysis, the Einstein single frequency (ω_e) model is used. The Einstein frequency is given by:

$$\omega_e = \sqrt{(g/M)} \tag{10.1}$$

where g = shear force constant, and M = effective atomic mass. The forces and masses can separated by forming the logarithm:

$$\ln \omega_e = (\ln g - \ln M)/2 \tag{10.2}$$

The thermal energy of an Einstein oscillator is $k\theta_e$ where k = Boltzman's constant, and θ_e is the "Einstein temperature." The mechanical energy of the oscillator is $h\theta_e/2\pi$ where h = Planck's constant.

Then an entropic characteristic temperature can be defined:

$$\ln(k\theta_s) = \ln(h\theta_e/2\pi) \tag{10.3}$$

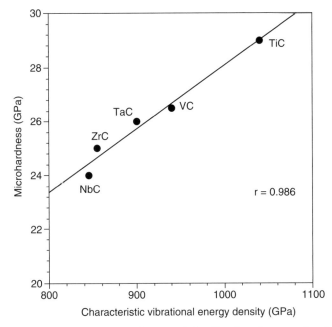

Figure 10.3   Carbide hardnesses vs. characteristic vibrational energy densities derived from average force constants (entropic specific heat). After Grimvall and Theissen (1986). The crystal structures are of the NaCi type. The hardness data are from Teter (1998).

And, using Equation 9.1, an effective force constant, g*, can be obtained:

$$g^* = (1/M)(h/2\pi k\theta_s) \qquad (10.4)$$

Entropy versus temperature data give values for θ_s, so values for g* can be obtained from Equation 10.3. These values depend on valence electron densities just as the elastic stiffnesses do.

To rationalize the units, g* is divided by the lattice parameter, a of each carbide. The final parameter (g*/a) = characteristic vibrational energy density has the units of energy per volume (GPa) which is the same as the hardness units. The correlation of this with hardness is shown in Figure 10.3. The correlation is good; especially when it is considered that the hardness numbers for carbides scatter as much as 30 percent.

10.3   TUNGSTEN CARBIDE

Tungsten carbide is of special interest because it retains its hardness to a high temperature compared with several other carbides. Thus titanium carbide is much harder (about 3200 VHN) at room temperature (compared with

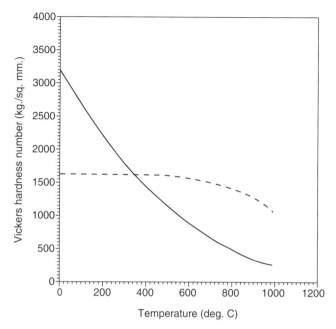

Figure 10.4   Comparison of the hot hardnesses of TiC and WC showing the crossover at about 350 °C.

1630 VHN) but at 800 °C, its hardness is about 1400 VHN, whereas TiC's hardness has dropped to only about 450 VHN (Figure 10.4).

Tungsten carbide is also of interest because its crystal structure is related to both the NaCl structure of TiC, and the AlB_2 structure of TiB_2. The structure consists of close-packed hexagonal W-atom layers stacked in a simple hexagonal structure with the carbon atoms occupying one-half of the hexagonal prismatic interstices. Therefore, the carbon atoms also form hexagonal layers between the tungsten layers (Figure 10.5).

The relation of the WC structure to the TiB_2 structure is that in the latter boron atoms occupy all of the prismatic interstices instead of half of them. The relation to the TiC structure is that the prismatic interstices become octahedral interstices in the TiC structure and there are half as many.

The hardness of WC is associated with the fact that the array of W-atoms in the cores of glide dislocations changes from hexagonal prismatic to quasi-octahedral so the coordination number of the C-atoms changes from approximately six to approximately eight. This increases the local electron density so dislocation motion is resisted.

Conjugate behavior occurs in TiC. In this case, at the cores of glide dislocations, the octahedral array of Ti-atoms changes to approximately hexagonal prismatic, so the coordination number of the C-atoms changes from eight to six.

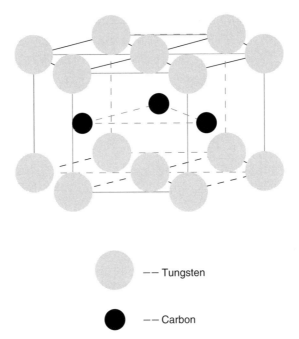

- - Tungsten

- - Carbon

Figure 10.5   Crystal structure of WC.

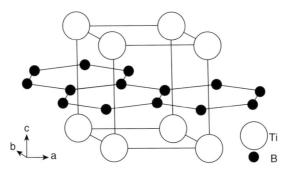

Figure 10.6   Schematic crystal structure of TiB_2.

10.4   BORIDES

The hardest of the transition-metal borides are the diborides. Their character-istic crystal structure (Figure 10.6) consists of plane layers of close-packed metal atoms separated by plane openly-patterned layers of boron atoms ("chicken-wire" pattern). If the metal atoms in the hexagonal close-packed layer have a spacing, d, then the boron atoms have a spacing of $d/\sqrt{3}$.

The B–B bond lengths in borides is close to that in pure boron crystals, and the latter are quite hard ($\approx 3000\,kg/mm^2$). Furthermore, the relative bond lengths in the borides are different from the carbides. For example, in TiB_2 the

TABLE 10.1   Vickers Hardness of Prototype Diborides

	H_v	Refer.
TiB_2	3350	(a)
ZrB_2	2300	(b)
HfB_2	2800	(b)
VB_2	2100	(a)
NbB_2	2600	(a)
TaB_2	2500	(a)

a. W. G. Fahrenholz et al., Jour. Amer. Chem, Soc., **90**, 1347 (2007).
b. A. A. Ivanko, *Handbook of Hardness Data*, Edited by G. V. Samsonov, translated from the Russian by Ch. Nisenbaum, Israel Program for Scientific Translations (available from U.S. Department of Commerce National Technical Information Service, Springfield, VA (1971).

B–B distance is 1.746 Å. In pure B, it is 1.75 Å. Therefore, covalent B–B bonds may be expected. During the complex deformation in an indentation, these strong bonds must be broken. They are the principal barriers to dislocation kink motion in the diborides.

Hardness values for the prototype diborides are listed in Table 10.1. Most hardness measurements for diborides have been made for sintered specimens; thus, they vary from one author to another. The values listed are the highest ones reported. Average values have little meaning in this case.

Perhaps the best known diboride is TiB_2, so it will be discussed in somewhat more detail.

10.5   TITANIUM DIBORIDE

To demonstrate the simple chemistry of the diboride structure, consider TiB_2, as an example (Figure 10.6). In Ti metal, the spacing of the atoms is 2.89 Å. In the diboride, the spacing is 3.03 Å, indicating some transfer of electrons into the metal. In pure boron the spacing is 1.76 Å, whereas it is 1.75 Å in the diboride. Thus the boron atoms of the hexagonal net pattern are covalently bonded. Overall, the diboride structure consists of interlamellar layers of metal and boron. This accounts for it having a combination of metallic (electrical conduction) and semiconductor (hardness) properties. Also, it suggests that plasmons in the boron layers influence the cohesion by interacting with plasmons in the titanium layers.

It has been found by Will (2004) from X-ray scattering measurements that valence electrons concentrate along the lines connecting the boron atoms, confirming that the boron layer is a covalently bonded network. The titanium layers are metallic. However. the layers are not characteristic of either pure Ti, or pure B, so the bonding is quite complex.

The mechanical behavior of TiB_2 is characterized by its lattice parameters, valence electron density, elasticity tensor, plasmon tensor, and its heat of formation. None of these probes the electronic structure directly, but taken together they indicate the expected magnitude of the hardness.

Although the primary glide plane of TiB_2 is the basal (0001) plane, indentation requires glide on more than one type of plane. Also, indentation is controlled by the most resistant deformation mode; in this case, probably the $(10\overline{1}0)\langle 10\overline{2}0\rangle$ mode. Therefore, the Burgers displacement, b, equals the length of the a-axis = 3.03 Å. The HOMO-LUMO gap of boron which measures bond strength is 1.55 eV. Equating this to the shear work done, $\approx \tau b^3/2$ during a displacement, b/2 yields $\tau = 18$ GPa, so H ≈ 36 GPa which is close to the measured value of 34 GPa. It may be concluded that the hardness of this and other diborides is determined by the strengths of their covalent chemical bonds.

10.6 RARE METAL DIBORIDES

The mechanical stabilities of solids are determined by their valence electron densities ($\#/cm^3$). A benchmark is provided by diamond. The volume of a carbon atom in diamond is about 5.68 Å^3, so the VED is 4/5.68 = 0.704 elec./Å^3. This is the highest of any element. Next highest is osmium whose volume is about 14 Å^3. Taking 2s + 6d for its valence electrons, this yields VED = 0.57 elec./Å^3. Adding B to make OsB_2 adds additional electrons, increases the volume, and increases the hardness. The VED becomes 0.51 elec./Å^3, which is a decrease compared with the metal. However, the hardness increases markedly from about VHN = 350 kg/mm^2 to 2500 kg/mm^2 (Cumberland et al., 2005).

The rare metal Rh lies adjacent to Os on the Periodic Table. Like Os, it forms a hexagonal diboride that is even harder. Its hardness is about 4800 kg/mm^2 (Chung et al., 2007), while its VED is about 0.477 elec./Å^3.

The crystal structures of these rare metal diborides are similar to Figure 10.6 except that the boride layers are puckered rather than flat.

10.7 HEXABORIDES

Figure 10.7 illustrates the prototype hexaboride crystal structure, that of lanthanum hexaboride. It consists of a simple cubic array of boron octahedra surrounding a metal atom at the body center of each cube. The octahedra are linked by B–B bonds connecting their corners. This makes the overall structure relatively hard with approximately the hardness of boron itself since plastic shear must break B–B bonds. The open volumes surrounded by boron octahedra are occupied by the relatively large lanthanum atoms as the figure shows schematically.

The hexaboride crystal structure is related to the CsCl structure so by analogy the glide planes are (100) and the glide directions are $\langle 100\rangle$. At the cores of glide disloca-tions the structure becomes quasi-hexagonal.

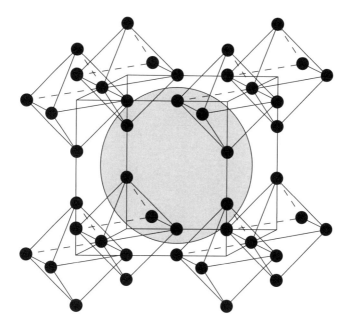

B–B Bond length = 1.75 Ang.
Lanthanum diameter = 3.73 Ang.

Figure 10.7 Crystal structure of Lanthanum Hexaboride (prototypre hexaboride). The black circles represent boron octahedra. They form a simple cubic arrangement surrounding the central metal atom.

An alternative version of the lanthanum hexaboride crystal structure has the boron octahedra occupying the body centered positions of the cubic array of lanthanum atoms (Figure 10.8). This version makes it clear that in order to plastically shear the structure, the boron octahedra must be sheared. Note that the octahedra are linked together both internally and externally by B–B bonds.

The hardnesses of some hexaborides are listed in Table 10.2.

Table 10.1 indicates that the hexaborides are quite hard. This hardness is associated with their strong bonds especially between the boron atoms, and a measure of it is the heat of formation. Consider the case of LaB_6. According to Topor and Kleppa (1984) the heat of formation is $\Delta H_f = -400\,kJ/mol. = 4.15\,eV$. The lattice parameter of the cubic unit cell is $a_0 = 4.156\,\text{Å}$, so the molecular volume is expected to be $71.8\,\text{Å}^3$. Then by the same arguments used for the carbides (Section 10.1), the hardness is expected to be $1850\,kg/mm^2$ which compares well with the observed value in Table 10.1.

The hardnesses of these crystals remains high up to the 600–700 C range and then decline rapidly (Chen, Xuan, and Otani, 2003). Also, it should be noted that the crystal hardnesses are anisotropic with the (110) surfaces more difficult to indent than the (100) surfaces (Li and Bradt, 1991).

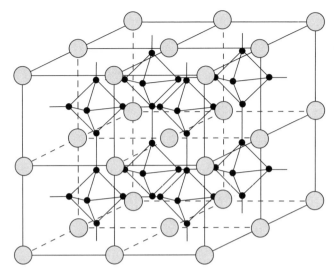

Figure 10.8 Alternative drawing of the crystal structure of Lanthanum Hexaboride with the metal atoms occupying the cube corners.

TABLE 10.2   Hardnesses of Some Hexaborides

Hexaboride	VHN (kg/mm^2)	Reference
Ca	2600	a
Sr	2900	b
Ba	3000	b
La	2000	c
Ce	1900	b
Sm	2000	b
Y	1800	c

a. S. K. Dutta, Amer. Cer. Soc. Bull., **54**, 727 (1975).
b. C. Miitterer et al., Surf. & Coatings Tech., **86–87**, 715 (1996).
c. C. S. Otani et al., Jour. Alloys & Compds., **350**, L4 (2003).

10.8   BORON CARBIDE (CARBON QUASI-HEXABORIDE)

Although the chemical formula for this compound is often written B_4C, crystallographic studies have shown that it is actually $B(CB_6)_2$, or alternatively, $B_{12}(CBC)$. That is, the structure contains 13 B atoms for every 2C, so it is nearly a hexaboride (Lee et al., 1991).

The hardness of boron carbide (carbon hexaboride) is not well defined because it is made as sintered compacts which have variable densities, compositions, and defect densities. It is very hard (up to 4400 kg/mm^2), and of relatively low density, so it has been used extensively as body-armor (McColm,

1990). Its hardness is associated with its B–C bonds (1.64 Å) and the C–B–C group (stronger than pure B–B bonds; l = 1.75 Å).

10.9   NITRIDES

Nitrides are closely related to carbides. Several of them have the same NaCl crystal structure, and similar lattice parameters. Also, the carbide and nitride of the same metal are mutually soluble. Their hardnesses are similar.

For example, consider the TiC and TiN pair. Their lattice parameters are 4.32 Å, and 4.23 Å, respectively; the difference is only two percent. Together with their mutual solubility (Schwarzkopf and Kieffer, 1953) this suggests that they have the same number of bonding valence electrons, although atomic carbon has four valence electrons, and atomic nitrogen has five. The extra nitrogen electron must be in a non-bonding state. This contradicts the valence electron concentrations assumed by Jhi et al., 1999.

Unfortunately, TiN is difficult to prepare in bulk form without porosity. Therefore, measured values of its properties show considerable scatter. In particular, consistent values for its hardness from one investigator to another are not available. They range from 1600 to 4000 kg/mm^2. For practical applications, TiN is used in the form of thin films. These are consistent in their hardnesses, but their complex structures make them difficult to interpret in terms of simpler quantities.

REFERENCES

A. H. Cottrell, *Chemical Bonding in Transition Metal Carbides*, Inst. Mater., London, UK (1995).

C. H. Chen, Y. Xuan, and S. Otani, "Temperature and Loading Time Dependence of LaB$_6$, YB$_6$ and TiC Single Crystals," Jour. Alloys and Compds., **350**, L4 (2003).

H. Y. Chung, M. B. Weinberger, J. B. Levine, A. Kavner, J. M. Yang, S. H. Tolbert, and R. B. Kaner, "Synthesis of Ultra-incompressible Superhard Rhenium Diboride at Ambient Pressure," Jour. Amer. Chem. Soc., **316**, 436 (2007).

R. W. Cumberland, M. B. Weinberger, J. J. Gilman, S. M. Clark, S. H. Tolbert, and R. B. Kaner, "Osmium Diboride, An Ultra-incompressible, Hard Material," Jour. Amer. Chem. Soc., **127**, 7264 (2005).

J. J. Gilman, "Hardnesses of Carbides and Other Refractory Hard Metals," J. Appl. Phys., **41**, 1664 (1970).

J. J. Gilman, *Electronic Basis of the Strength of Materials*, p. 115, Cambridge University Press, Cambridge, UK (2003).

G. Grimvall and M. Thiessen, "The Strength of Interatomic Forces," p. 61, in *Science of Hard Materials—Proc. Int. Conf., Rhodes—Inst. Phys. Conf Series #75*, Edited by E. A. Almond, C. A. Brookes, and R. Warren, Adam Hilger Ltd, Bristol, UK (1986).

D. He, Y. Zhao, L. Daemen, J. Qian, and T. D. Shen, "Boron Suboxide: As Hard as Cubic Boron Nitride," Appl. Phys. Lett., **81**, 643 (2002).

S.-H. Jhi, J. Ihm, S. G. Louie, and M. L. Cohen, Narure, **399**, 132 (1999).

H. Ledbetter, "Atomic Frequency and Elastic Constants," Zeit. Fur Metalikunde, **82**, 820 (1991).

S. Lee, S. W. Kim, D. M. Bylander, and L. Kleinman, "Crystal Structure. Formation Enthalpy, and Energy Bands of B_6O," Phys. Rev. B, **44**, 3550 (1991).

H. Li and R. C. Bradt, "Knoop Microhardness Anisotropy of Single-crystal LaB_6," Mater. Sci. Eng. A, **142**, 51 (1991).

I. J. McColm, *Ceramic Hardness*, p. 231, Plenum Press, London, UK (1990).

P. Schwarzkopf and R. Kieffer, *Refractory Hard Metals*, The Macmillan Company, New York, USA (1953).

P. S. Spoor, J. D. Maynard, M. J. Pan, D. J. Green, J. R. Hellman, and T. Tanaka, "Elastic Constants and Crystal Anisotropy of Titanium Diboride," Appi. Phys. Lett. **70**, 1959 (1997).

D. M. Teter, "Computational Alchemy: The Search for New Superhard Materials," Mater. Res. Soc. Bull., **23**, 22 (1998).

L. Topor and O. J. Kleppa, "Standard Molar Enthalpy of Formation of LaB_6 by High-Temperature Calorimetry," Jour, of Chem. Thermo., **16**, 993 (1984).

G. Will, "Electron Deformation Density in Titanium Diboride: Chemical Bonding in TiB_2," Jour. Sol. St. Chem., **177**, 628 (2004).

11 Oxides

11.1 INTRODUCTION

Oxide crystals are the most common of all crystals because oxygen combines readily with almost all metals and semiconductors. It combines particularly strongly with silicon which is very abundant, constituting about 26 percent of the Earth's crust. The combination forms silicate tetrahedra and they link together in a great variety of ways, including chains, sheets, and frameworks. These are the basis of a vast variety of natural and synthetic minerals. Most of these are too complex mechanically for simple analysis, so they will not be discussed here. Discussion here will be limited to two examples, quartz and talc. The former is a prototype for hard silicates, and the latter is the softest material considered by the Mohs scale. Other oxides, including alkaline earths, borates, garnets, and perovskites, will also be discussed.

The number of oxide type minerals is quite large. Kostov (1956) has identified 160 specific minerals, grouped them into classes (chrysoberyl, spinel, corundum, periclase, etc.), and proposed a classification system. Only a few examples will be discussed here.

Methods for relating hardness values to other physical properties are presented, particularly chemical bond strengths. There is no universal method for doing this, although there have been attempts by other authors to do it with varying degrees of success. For example, see Gao, 2004.

11.2 SILICATES

The silicates are a large class of solids of great importance in industry as well as science, particularly geology. The prototype silicate is quartz consisting of SiO_4 tetrahedra which share their corners and edges and are arrayed in various three-dimensional patterns depending on the temperature. In other crystalline minerals the tetrahedra are linked in one-dimensional chains, or two-dimensional sheets. The arrays in these latter cases are combined with various metal ions.

Chemistry and Physics of Mechanical Hardness, by John J. Gilman
Copyright © 2009 John Wiley & Sons, Inc.

In addition to the large number of silicate crystals, the SiO_4 tetrahedra are the basis of many glasses in which the structure derives from that of the liquid state. These structures have short-range but not long-range order.

11.2.1   Quartz

An apparent anomaly of quartz and related silicates is that SI–O bonds are very strong (3.8 eV, compared with C–C at 3.6 eV), but quartz is not especially hard. On the 10-level Mohs scale it is only about seven (diamond is 10). Viewed another way, the energy gap between the bonding states and the anti-bonding states in quartz is about eight eV which is significantly larger than diamond at 5.7 eV, but diamond is much harder. This anomaly is resolved by considering the structures of dislocation cores in quartz. Also, it illustrates why the bond modulus is a more effective parameter than the gap itself in determining hardness (Gilman, 2007).

Figure 11.1 illustrates the crystal structure of quartz, comparing it with silicon (diamond structure). The form of quartz in the figure is that of cubic crystobalite, its most simple form at elevated temperatures. This form is easier to illustrate than the quartzes of lower symmetry. Also, although the structures of both Si and crystobalite are usually displayed using a cubic unit cell, a tetragonal cell is used here because it visualizes the atomic relationships better. The two images in the figure are scaled to match the observed scales for Si and SiO_4. It may be seen that the chrstobalite structure is derived from the silicon structure by placing an oxygen atom at the center of each Si–Si bond, and then adjusting the bond lengths to the values shown in the figure (Bragg et al., 1965).

As discussed in Chapter 4, one way of viewing a dislocation is that it is a microscopic lever that concentrates stress. Levers transduce mechanical work (force x distance) from large forces moving short distances to small forces moving large distances. Thus dislocations convert small macroscopic stresses into large microscopic stresses. As a result, not just an amount of work (energy) is important, but also the length scale that is associated with it.

Dislocation lines do not move concertedly, that is, all at once. They move, by forming "kinks" along their lengths, and when the kinks move, the lines move. The open crystal structure of quartz (crystobalite) results in a relatively large amount of volume being associated with a kink on a dislocation line. This relatively large volume lowers the value of quartz's bond modulus, making its hardness consistent with those of other covalently bonded substances.

The distance that the small segment of a dislocation line moves when a kink moves is called the Burgers displacement, b. Figure 11.2 illustrates it for the case of quartz. It determines the amount of work that is done by the advance of a kink (per unit width of the kink) which is acted upon by the virtual force generated by the applied shear stress, τ. This force is τb per unit length of the dislocation line. Letting the kink width be b since the displacement is b, the work done is τb^3. This is resisted by the strength, U (eV) of a Si–O bond which

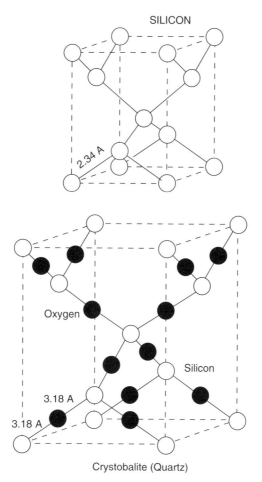

Figure 11.1 Comparison of the atomic structure of cristobalite (high temperature form of SiO_2) with that of silicon (diamond structure using tetrahedral unit cell).

must be sheared to allow the kink to advance. Equating the driving work and the resisting energy gives a value for the required stress, $\tau \approx U/b^3$. Here b^3 is approximately a molecular volume. This is consistent with other covalently bonded substances.

Numerically, the Burgers displacement which spans two SiO_2 tetrahedra is 4.28 Å. Therefore, the pertinent volume of a kink is approximately $b^3 = 78.4 \, \text{Å}^3$ which is large enough to considerably reduce the bond modulus (energy gap/ molecular volume). With $E_g = 8 \, \text{eV.}$, BM $= 0.102 \, \text{eV/Å}^3$, or 10.2 GPa. This compares well with the hardness of quartz, 12 GPa (1200 kg/mm^2).

The hardness (480 kg/mm^2) of the quartz version of GeO_2 is also consistent with its bond modulus. Unfortunately, SnO_2 does not have a quartz like structure so there are only two members of this isoelectronic set (SiO_2 and GeO_2).

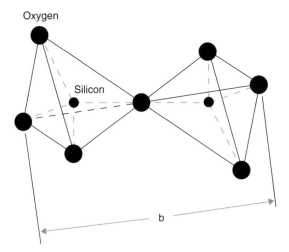

Figure 11.2   Burgers displacement in quartz.

11.2.2   Hydrolytic Catalysis

Geologists have found that the mineral quartzite as well as synthetic quartz is softened by water (Griggs and Blacic, 1965). Dry quartz crystals retain their hardnesses to nearly 1000 °C, but are weakened considerably (nearly an order of magnitude) by the presence of water. Similar weakening is observed for olivine $(Mg,Fe)_2(SiO_4)$ and feldspar $(KAlSi_3O_8)$.

It seems that water hydrolyzes the Si–O–Si connections between the slica tetrahedra, yielding Si–OH:HO–Si. That is, strong –O– bridges are replaced by weak H:H hydrogen bridges. These become associated with kinks on dislocation lines increasing the mobilities of kinks and therefore dislocations (Griggs, 1967). Since the concentration of kinks is small compared with the total number of atoms, relatively little water is needed for this catalytic mechanism.

11.2.3   Talc

A bar of talc feels like a bar of soap which is why it is often called "soapstone." Its exceptional softness (it is the softest of the Mohs minerals) is a direct result of its unusual crystal structure. This consists of sheets of silicate tetrahedra without metal ions between the sheets. Thus the sheets are bonded only by London polarization forces. The latter are particularly weak because silicate tetrahedra have relatively small polarizabilities.

Talc is a hydrated magnesium silicate, $Mg_6(Si_8O_{20})(OH)_4$. It is a layerd compound like mica. One layer of its crystal structure is shown schematically in Figure 11.3. Such layers are stacked up like playing cards in real crystals. Notice that the top and bottom of the layer consist of slicate tetrahedra with oxygen

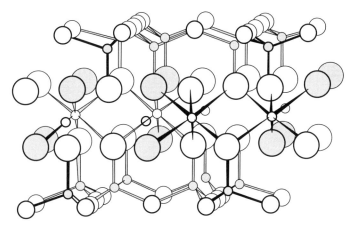

Figure 11.3   Schematic crystal structure of one layer of talc. Such layers are stacked to make the complete structure. The large and medium open circles represent oxygen atoms. The cross-hatched large circles represent hydroxyls (OH). The small open circles represent magnesium atoms (Mg); and the cross-hatched small circles represent silicon (Si) atoms. Figure is reproduced from Bragg, Claringbull and Taylor (1965).

atoms at the very top and bottom. Thus the layers have no chemical bonds between them. This is why talc shears so easily.

Mica and other layered minerals differ from talc because metal atoms lie between their layers producing some chemical bonding. Also, their layers are usually stronger because Al replaces (partially or fully) the central Mg layer of Figure 11.3.

11.3   CUBIC OXIDES

11.3.1   Alkaline Earth Oxides

Perhaps the most simple crystals in this class are the alkaline earth oxides. They are II–VI compounds and have rocksalt crystal structures. Data for their hardnesses versus their bond moduli (optical band gaps per molecular volumes) are displayed in Figure 11.4.

A similar linear dependence is found for their reciprocal molecular volumes which are proportional to their polarizabilities, α. Thus H is expected to be proportional to $1/\alpha$ and indeed it is (Dimitrov and Komatsu, 2002). See Figure 11.5. Furthermore, since their shear moduli C_{44} are proportional to $1/\alpha$ (Gilman, 1997), the graph also indicates that their hardnesses are proportional to their shear moduli (Singh et al., 2007).

In addition, it has been found that the hardnesses of these simple oxides are proportional to their heat of formation densities (that is, their heats of formation divided by their molecular volumes). Thus, the hardnesses of these

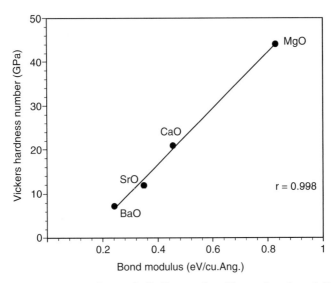

Figure 11.4   Hardness of alkaline earth oxides vs. bond moduli.

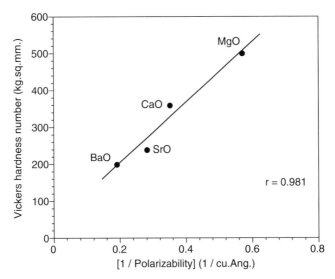

Figure 11.5   Same as Figure 11.4 with H versus reciprocal polarizabilties.

compounds are consistent with their other physical properties and with the idea that chemical bonding strengths determine hardness.

11.3.2   Perovskites

Perovskites are compounds of the ABC_3- type where C is often oxygen, but not always. Figure 11.6 shows two versions of the perovskite crystal structure

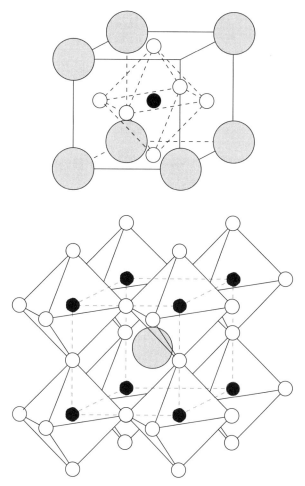

Figure 11.6 Views of perovskite crystal structure. Top—conventional cubic unit cell: white circles = oxygen; black circle = transition metal; gray circles = alkali or alkaline earth metal. Bottom—extended unit cell to show the cage formed by the oxygen octahedra. Adapted from Bragg et al. (1965).

of the prototype $CaTiO_3$ compound. The given cubic structures are ideal ones. Often the real structures have small deviations from cubic ones. The top schematic image shows the "textbook" unit cell, while the bottom image is intended to clarify the structure by showing the octahedral cages explicitly in an extended unit cell. In both cases, the large gray circles represent the Ca ions; the black circles represent the Ti ions; and the open circles represent oxygen.

The most abundant of all minerals in the interior of the earth is $(Mg,Fe)SiO_3$ perovskite. It constitutes greater than seventy percent of the lower mantle, so it is of great importance to geophysics. At room temperature the hardness of $MgSiO_3$ is VHN = 1800 kg/mm^2 and its Chin-Gilman parameter is 0.01.

TABLE 11.1   Cubic Perovskites

	G (GPa)	VHN (GPa)	VHN/G
$BaUO_3$	46	5.5	.12
$BaZrO_3$	103	5.0	.05
$BaMoO_3$	94	3.2	.03
$SrTiO_3$	99	7.8	.08
$SrMoO_3$	70	5.5	.08

The latter indicates that the dominant bonding type is covalent. This was also observed for $CaTiO_3$ and $BaTiO_3$, both of which have the perovskite crystal structure, but are considerably softer than $MgSiO_3$. The Mg perovskite is about twice as hard as crystobalite (quartz). However, hydration converts $MgSiO_3$ to talc, which is very soft.

The perovskite crystal structure is exhibited by a large number of compounds because numerous metals can form the octahedral sub-units, and other metal ions can lie between the sets of eight octahedra. The main constraint is on the sizes of the ions that can be chosen to fit compactly together.

It may be apparent from studying the perovskite structure that it is likely to exhibit quite anisotropic plastic (hardness) behavior, and it does. The primary glide plane is (110) and the glide direction is $\langle 1\text{--}10 \rangle$.

Some perovskites are widely used as piezo-transducers, $BaTiO_3$ for example, and lead zirconate ($PbZrO_3$) which is a well-known ferroelectric material sensitive to stresses. Also, some perovskites are good pyro-transducers; that is, heat causes electric polarization of them.

If Fe atoms occupy the "B" sites (gray in the figure), they may be ferromagnetic, and known as ferrites, although many of the latter have a different structure, that of magnetite.

Electronegative metals such as Na in the "B" sites may lead to high electrical conductivity as in the tungsten "bronzes" ($NaWO_3$).

The hardnesses of some perovskites are given in Table 11.1 (based on the data of Yamanaka et al., 2004). The table shows that these perovskites are moderately hard and the third column which lists their Chin-Gilman parameters indicates that they are predominately ionically bound.

The dependence of hardness on the valence electron densities is illustrated by Figure 11.7.

11.3.3   Garnets

Garnets are important gems, abrasives, microwave systems components, magnetic bubble memories, and laser hosts. For the latter, yttrium aluminum garnet is the most important. It also plays an important role in aircraft turbines where it forms a protective coating on the turbine blades.

The garnet structure has high overall symmetry (cubic) but a complex structure (Bragg and et al., 1965). The prototype is the mineral Grossularite

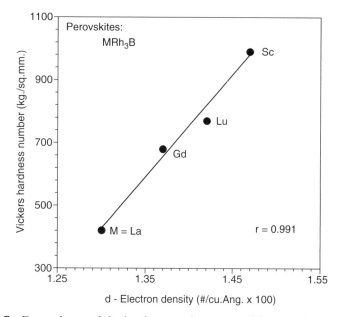

Figure 11.7 Dependence of the hardnesses of some transition metal-rhodium-boron perovskites on their d-electron densities.

($Ca_3 Al_2 Si_3 O_{12}$) in which silica tetrahedra and aluminum octahedra are linked together so that sets of eight oxygen ions form quasi-cubic cells with calcium ions at their centers (Figure 11.8). The overall cubic unit-cell contains eight formula units (80 atoms) and has a cell parameter of 11.82 Å.

11.3.3.1 ($Y_3Al_5O_{12}$)—YAG Most garnets are silicates, whereas yttrium aluminum garnet (YAG) is an aluminate. In YAG, both the tetrahedral and the octahedral holes of the garnet structure are occupied by Al-ions and the quasi-cubic holes are occupied by Y-ions.

At room temperature the hardness of YAG is about VHN = 1700 kg/mm². This is considerably harder than YIG (yttrium iron garnet) at 1200 kg/mm², or GGG (gadolinium gallium garnet) at 1350 kg/mm² (Sirdeshmukh et al., 2001). However, the outstanding aspect of the hardness of YAG is its persistence at high temperatures. It is the hardest of all oxides at 1300 °C, about 850, compared with 600 for chrysoberyl ($BeAl_2O_4$), and 250 for sapphire (Al_2O_3), all in kg/mm², (Gilman, 2004). See Figure 15.1.

The intrinsic energy band-gap of YAG is about 6.6 eV., and the Burgers displacement is about half the unit cell size, or 6 Å. Then, if a kink volume is taken to be $6 \times 3 \times 3 = 54$ Å³, the bond modulus is 0.11 eV/Å³, or 1800 kg/mm². Given how little is known about dislocation motion in garnet, this agreement with the room temperature hardness value is largely fortuitous.

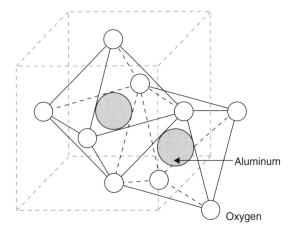

Figure 11.8 Shared faces of oxygen octahedra in the Al_2O_3 crystal structure and related oxides. This generates the chemical stoichiometry and places pairs of Al atoms in close proximity.

11.4   HEXAGONAL (RHOMBOHEDRAL) OXIDES

In most oxides, the oxygen atoms are present as close-packed layers stacked either to produce cubic symmetry or hexagonal symmetry. Some of the cubic cases have already been discussed. Now some hexagonal cases will be considered.

11.4.1   Aluminum Oxide (Sapphire)

The crystal structure of Al_2O_3 consists of approximately close-packed oxygen atoms stacked in an ABABAB type of sequence (Bragg et al., 1965, p.96). Between each pair of layers, two-thirds of the octahedral interstices are occupied by Al atoms. As a result the pattern repeats after every six layers of oxygen, so the c-axis is 13 Å long. Another feature of the structure is that groups of three oxygens form common faces of two adjacent octahedra. In each of these subgroups two Al atoms are linked as illustrated in Figure 11.9. The Al–Al distance is about 2.76 Å which is somewhat smaller than the spacing in the metal, 2.86 Å, or about 3.5 percent smaller.

At low temperatures, Al_2O_3 is hard and brittle, but it can be plastically deformed at high temperatures. The primary glide plane is the basal (0001) plane, and the Burgers displacement at low temperatures is 5.84 Å. When the Al atoms become mobile at high temperatures this shortens to about 2.76 Å.

The hardness of Al_2O_3 is VHN = 2700 kg/mm^2 and its rms. shear stiffness is 366 GPa so its Chin-Gilman parameter is 0.074. This suggests that its chemical bonding is a combination of covalent and ionic bonding.

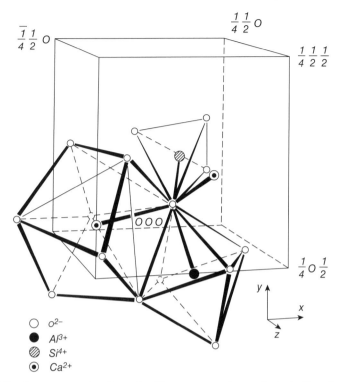

Figure 11.9 Arrangementt of ions in silicate garnet (grossularite). Showing tetrahedral, octahedral, and quasi-cubic groups. After Geller (1967).

In order for a kink on a dislocation line to move it must shear (destroy) Al_2O_3 subunits of the crystal structure. This requires approximately the heat of formation, ΔH_f of Al_2O_3 which is 402 kcal/mol = 17 eV/molecule (Roth et al., 1940). The work done by the applied shear stress must supply this energy. This is about τb^3 so the shear stress required is about 13.7 GPa, and the hardness, H, is about twice this, or 27.4 GPa, which is close to the observed hardness of 27 GPa.

It may be concluded that the hardness of Al_2O_3 is determined by the strength of its chemical bonds. This is probably also the case for other A_2B_3 oxides, such as those with A = Fe, Cr, Ti, Nb, Y, etc.

11.4.2   Hexaboron Oxide

Because of its relatively loosely bound outermost electron, and its small size, boron reacts readily with other atoms (including itself) to form a variety of crystal structures. The B–B bond is only about 12 percent longer (1.75 Å) than the very strong C–C bond (1.54 Å). It not only forms very hard compounds with carbon and nitrogen, but also with oxygen.

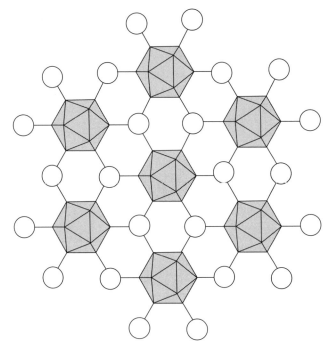

Figure 11.10 Crystal structure of oxygen hexaboride showing one layer of the hexagonal structure.

The structure of oxygen hexaboride illustrates the flexibility of boron. In this case, the boron atoms form iscosahedra (12 faces) instead of octahedra (eight faces). This requires a rhombohedral (hexagonal) array of the oxygen atoms and the icosahedra. Figure 11.10 shows the arrangement in one layer of this structure. Two other layers have the same form but are rotated 60° relative to one another, and the fourth layer repeats the position of the first layer.

Good quality OB_6 crystals can be grown at high pressures and temperatures from a flux. For an indenter loading force of 1N, their VHN is 4500 kg/mm^2 which is only six percent less than the 4800 kg/mm^2 measured by the same authors for cubic BN. It is about 30 percent greater than values measured for sintered compacts of polycrystalline oxygen hexaboride.

The oxygen atoms lie about 3.1 Å apart. This is much larger than the 1.3 Å in free oxygen molecules, so there is essentially no bonding between the oxygen atoms. On the other hand, the O–B separations are only 1.43 Å. Using this as the diameter of the bond, volume = 1.53 Å^3, and five valence electrons (three for B and two for O) the VED is about 1.96 elect./A^3 which is only about five percent less than the VED (2.07 elect./A^3) of the C–C bond. This may account for the great hardness of OB_6.

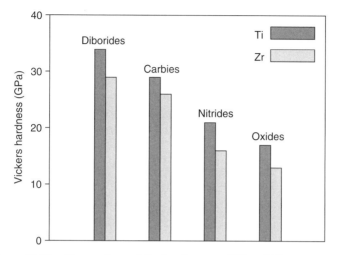

Figure 11.11   Comparison of the hardnesses of Ti and Zr compounds.

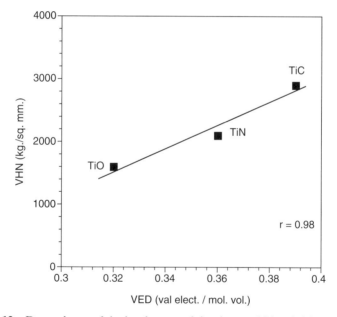

Figure 11.12   Dependence of the hardnesses of titanium carbide, nitride, and oxide on their valence electron densities (VEDs).

11.5   COMPARISON OF TRANSITION METAL OXIDES WITH "HARD METALS"

In the series TiB_2, TiC, TiN, TiO, the hardness gets progressively smaller. This is illustrated in Figure 11.11 for two of the transition metals, Ti and Zr, but is also a trend for other ones. The trend is related to the decreasing electro-negativity of the non-metal atoms which decreases their tendencies to transfer electrons to the metal atoms in the various compounds.

Three of these compounds have cubic symmetry, while TiB_2 has hexagonal symmetry. Since they are metallic, bond moduli cannot be defined for them, but valence electron densities can be. The hardnesses of the cubic titanium compounds depend linearly on their VEDs; the numbers of valence electrons are $(4 + 4 = 8)TiC$, $(4 + 3 = 7)TiN$, and $(4 + 2 = 6)TiO$. The linear dependence is shown in Figure 11.10. A similar linear dependence on their $C_{44}s$ is also found (Figure 11.12).

REFERENCES

Sir L. Bragg, G. F. Claringbull, and W. H. Taylor, *Crystal Structures of Minerals—Vol. IV*, p. 276, Cornell University Press, Ithaca, NY, USA (1965).

V. Dimitrov and T. Komatsu, "Classification of Simple Oxides: A Polarizability Approach," Jour. Sol. St. Chem., **163**, 100 (2002).

F. Gao, "Hardness Estimation of Complex Oxide Materials," Phys Rev. B, **69**, 094113 (2004).

S. Geller, "Crystal Chemistry of the Garnets," Zeit. Kristallographie, **125**, 1 (1967).

J. J. Gilman, "Chemical and Physical 'Hardness,'" Mater. Res. Innov., **1**, 71 (1997).

J. J. Gilman, "Monocrystals Offer Best Route to Ultrahigh-Strength Materials," MRS Bulletin, p.678, October (2004).

J. J. Gilman, "Bond modulus and stability of covalent solids," Phil. Mag. Lett., **87**, 121 (2007).

D. T. Griggs and J. D. Blacic, "Quartz: Anomalous Weakness of Synthetic Quartz Crystals," Science, **147**, 292 (1965).

D. T. Griggs, "Hydrolytic Weakening of Quartz and other Silicates," Geophysical Jour. Of the Royal Astronomical Soc., **14**, 19 (1967).

I. Kostov, "The Classification of Oxides," Chapter 14 in *Aspects of Theoretical Mineralogy in the USSR*, Trans. by M. H. Battey and S. I. Tomkeieff, The Macmillan Company, New York, USA (1956).

W. A. Roth, U. Wolf, and O. Fritz, "The Heat of Formation of Aluminum Oxide (Corundum) and of Lanthanum Oxide," Zeit. Elektro. u. Angew. Physik. Chemie, **46**, 42 (1940).

S. P. Singh, S. Gupta, and S. C. Goyal, "Elastic Properties of Alkaline Earth Oxides under High Pressure," Physica B, **391**, 307 (2007).

D. B. Sirdeshmukh, L. Sirdeshumkh, K. G. Subhadra, K. Kishan Rao, S. Bal Laxman, "Systematic Hardness Measurements on Some Rare Earth Garnet Crystals," Bull. Mater. Science, **24**, 469 (2001).

S. Yamanaka, K. Kurosaki, T. Maekawa, T. Matsuda, S. Kobayashi, and M. Uno, "Thermochemical and Thermophysical Properties of Alkaline-earth Perovskites," Jour. Nucl. Mater., **344**, 61 (2004).

12 Molecular Crystals

12.1 INTRODUCTION

In molecular crystals, there are two levels of bonding: intra—within the molecules, and inter—between the molecules. The former is usually covalent or ionic, while the latter results from photons being exchanged between molecules (or atoms) rather than electrons, as in the case of covalent bonds. The hardnesses of these crystals is determined by the latter. The first quantum mechanical theory of these forces was developed by London so they are known as London forces (they are also called Van der Waals, dispersion, or dipole-dipole forces).

There are two general cases of dipole-dipole forces: those between molecules in which the distribution of electronic charge is centrosymmetric and those in which it is not. In the first case, there are no permanent electrical dipoles, whereas there is a permanent dipole if the charge distribution is non-centro-symmetric. When permanent dipoles are not present, there are nevertheless fluctuating dipoles as a result of atomic vibrations. These are always present because of zero-point motion. At temperatures greater than $0\,°K$, thermal energy further excites the molecular vibrational modes which create fluctuating electric dipoles.

As molecular dipoles vibrate, they emit photons which excite vibrations in nearby molecules. In turn, these molecules emit photons which interact with the initiating molecule. In this way, the molecules interact by exchanging photons. Again there are two modes. In one case, the vibrations of the molecules occur in phase with one another. In the second case, they interact out of phase. The energy of the system is lower when the vibrations are in phase, so this case creates attractions between the molecules, while the out-of-phase case creates repulsions. Since the energy of the in phase case is lower, the net effect is attraction.

Dipole-dipole forces are weaker than electrostatic forces, but they can represent a substantial fraction of monopole forces. They have important effects because they are predominantly positive. Therefore, they add up, and even though they decay rapidly with the distance between molecules, their sums remain significant, leading to measurable adhesive forces between macroscopic solid bodies.

Chemistry and Physics of Mechanical Hardness, by John J. Gilman
Copyright © 2009 John Wiley & Sons, Inc.

The term "molecular crystal" refers to crystals consisting of neutral atomic particles. Thus they include the rare gases: He, Ne, Ar, Kr, Xe, and Rn. However, most of them consist of molecules with up to about 100 atoms bound internally by covalent bonds. The dipole interactions that bond them is discussed briefly in Chapter 3, and at length in books such as Parsegian (2006). This book also discusses the Lifshitz-Casimir effect which causes macroscopic solids to attract one another weakly as a result of fluctuating atomic dipoles. Since dipole-dipole forces are almost always positive (unlike monopole forces) they add up to create measurable attractions between macroscopic bodies. However, they decrease rapidly as any two molecules are separated. A detailed history of intermolecular forces is given by Rowlinson (2002).

Molecules can link together in one dimension to form chains (threads), or in two dimensions to form membranes, or in three dimensions to form blocks. Hardness has meaning only for the last case. Long chains of molecules constitute polymers and will be discussed in the next chapter. Small molecules formed of a few atoms are gases or liquids at room temperature, so hardness has no meaning for them.

The world of molecular crystals is vast, and because of their weak bonds their hardnesses are relatively small, so only a few of them will be discussed here. For a more comprehensive discussion of these materials, the reader may see the book by Schwoerer and Wolf, 2007.

There is a rough correlation between the hardnesses and the cohesive energies of molecular crystals as shown by Roberts et al. (1995). These authors studied crystals of 11 pharmaceutical compounds and found a linear correlation between their hardnesses and their cohesive energies. However, the data scatter substantially. The hardnesses range from about 1.0 (aspirin), through 5.0 (sucrose), to 10.0 (anthracene) kg/mm^2.

Trends in the hardnesses of molecular crystals are similar to those of inorganic crystals (Stephens et al., 2003). Thus, mixed crystals are harder than their pure components, crystals with foreign solutes are harder than pure crystals, and molecular crystals may be anisotropic.

12.2 ANTHRACENE

Aromatic compounds in the series, benzene, napthalene, anthracene, tetracene, etc., form crystals. However, benzene melts below room temperature. Napthalene, although solid at room temperature, has a high vapor pressure. Therefore, the first in the series whose crystals are stable enough at room temperature for extensive hardness studies is anthracene.

Molecules of anthracene consist of three shared quasi-hexagonal rings of carbon (14 C-atoms) plus ten H-atoms attached to ten of the carbons (Figure 12.1). In three dimensions the anthracene molecules have the shape of elongated tablets. They stack in crystals side-by-side in a staggered pattern to form sheets which pack together in herringbone arrays.

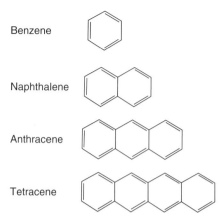

Figure 12.1   Schematic series of aromatic polyacene molecules.

The hardnesses of the resulting crystals are relatively small, about $10 \, kg/mm^2$ (Vaidya et al., 1997). Also, as might be expected from their layered structures, their hardnesses are quite anisotropic (Sasaki and Iwata, 1984).

Kojima (1981) discovered that a photo-plastic effect occurs in anthracene. It is largest for light of 430 nm. wavelength and is partially reversible. The effect probably results from a change in the polarizabilities of the anthracene molecules caused by photo-excitation. This is expected to increase the cohesion in the crystals slightly. The magnitude of the effect is up to about ten percent.

Anthracene is hardened by about a factor of two by dissolving phenanthrene in it (Vaidya and Shah, 2003), but the authors do not state the solute concentration.

12.3   SUCROSE

The molecular weight of sucrose ($C_{18}H_{32}O_{16}$) is about $504 \, g/mol$ so it is a relatively large but compact molecule. Crystals of it are aggregates of these globular molecules held together by London forces. They are brittle, but small plastic indentations can be made in them to yield an average hardness number of about $160 \, kg/mm^2$ (Ramos and Bahr, 2007).

12.4   AMINO ACIDS

Amino acid molecules consist of an amino group (NH_2) combined with an acid carboxyl group (COOH), and an H atom, as well as a residual group, R. In biological systems there are twenty different R groups including aliphatic chains, aromatic loops, and some containing sulfur plus hydrocarbons. The most simple is glycine where R = H (Stryer, 1988).

Hardness data for only two amino acids were found in the literature. They are glycine and alanine. They are the smallest of the amino acids. Both consist of rather flat tablet-like collections of atoms that form layered crystal structures in which the molecular sub-groups within the layers are held together by hydrogen bonds (Albrecht and Corey, 1939), and the molecules by London forces. Their hardnesses are:

γ – glycine (H = 29 kg/mm^2) – K. Ambujan et al., Crys. Res. Tech., **41**, 671 (2006)

l – alanine (H = 120 kg/mm^2) – L. Misoguti et al., Opt. Mater., **6**, 147 (1966)

And their chemical compositions are:

γ – glycine = $C_2NO_2H_5$
l – alanine = $C_3NO_2H_7$

The hardness value reported for alanine seems rather high, whereas the glycine value is typical for molecular crystals.

12.5 PROTEIN CRYSTALS

Proteins are poly-amino acids. The monomer acids combine when the carboxyl group of one amino acid reacts with the amino group of a neighbor. The former loses an OH, and the latter an H. These become H_2O, and a "peptide bond" forms as a link between each monomer pair yielding a poly-peptide. Since there are twenty amino acids in biological systems which can join together in any sequence and the chains can be short or very long, the number of protein molecules is almost limitless. Monolithic poly-peptides may also form of almost any length.

The hardness of only one type of protein crystal has been found in the literature. It is for lysozyme. This is an enzyme found in egg whites and tears. It destroys bacterial membranes. It is relatively small for a protein molecule, consisting of a chain of 129 amino acids folded into a globule with the volume $\approx$ 30,000 Å^3. Its crystals are aggregates of these globular molecules held together by London forces (Stryer, 1988).

In an environment with high humidity lysozyme is quite soft with a hardness number of only VHN = 2 kg/mm^2 at 295 K (Koizumi et al., 2004). It softens at 305 K to 1 kg/mm^2 and hardens to 2.7 kg/mm^2 at 285 K.

Glide bands are observed around hardness indentations in lysozyme so dislocations (with large displacements) are associated with its deformation.

A shear modulus of about 1 GPa has been measured for wet lysozyme. Thus its Chin-Gilman parameter is about 0.02 which is large compared with metals and small compared with covalent crystals.

Table 12.1

Crystal	VHN (kg/mm^2)	Reference
NH$_4$ClO$_4$	13	Elban and Armstrong, 1998
RDX	24	Hagan and Chaudhri, 1977
PETN	18	Amuzu et al., 1976

12.6   ENERGETIC CRYSTALS (EXPLOSIVES)

An interesting class of molecular crystals are those that easily decompose exothermally; i.e., explosive crystals. Some form from small molecules such as lead azide [Pb(N$_3$)$_2$] and ammonium perchlorate (NH$_4$ClO$_4$) and others form moderately large molecules such as RDX (cyclotrimethylenetrinitramine), and PETN (pentaerythritol-tetranitrate).

The Vickers hardnesses of a few energetic crystals are given in Table 12.1.

Note that these hardness values are approximate because these crystals fracture very easily. The fracture surface energy of PETN is only 0.11 J/m^2 (110 erg/cm^2) (Hagan and Chaudhri, 1977).

The Chin-Gilman parameter for PETN where the shear modulus is known is about 0.036 which is consistent with other molecular crystals.

12.7   COMMENTARY

Molecular crystals come in too many varieties and mixtures of chemical binding for simple theories of their hardnesses to be feasible. This is aggravated by their relatively low symmetries, making them quite ansotropic. Rough estimates of their hardnesses can be made if their shear moduli are known using the Chin-Gilman parameter. However, the shear moduli have been measured in only a few cases.

In general, molecular crystals are too soft for them to be of interest as structural materials. Also, they fracture readily. Because of their transparencies and non-linear properties some of them are of interest for optical applications, but most of them suffer from optical damage at low intensities of light.

REFERENCES

G. Albrecht and R. B. Corey, "The Crystal Structure of Glycine," J. Amer. Chem. Soc., **61**, 1087 (1939).

J. K. A. Amuzu, B. J. Briscoe, and M. M. Chaudhri, "Frictional Properties of Explosives," Jour. Phys. D, **9**, 133 (1976).

W. L. Elban and R. W. Armstrong, "Plastic Anisotropy and Cracking at Hardness Impressions in Single Crystal Ammonium Perchlorate," Acta Mater., **46**, 6041 (1998).

J. T. Hagan and M. M. Chaudhri, "Fracture Surface Energies of High Explosives PETN and RDX," Jour. Mater. Sci., **12**, 1055 (1977).

K. Koizumi, M. Tachibana, H. Kawamoto, and K. Kojima, "Temperature Dependence of Microhardness of Tetragonal Hen-egg-white Lysozyme Single Crystals," Phil. Mag., **84**, 2961 (2004).

K. Kojima, "Photoplastic Effect in Anthracene Crystals," Appl. Phys. Lett., **38**, 530 (1981); see also, *ibid.*, **56**, 927 (1984).

V. A. Parsegian, *Van Der Waals Forces*, Cambridge University Press, New York, USA (2006).

K. JU. Ramos and D. F. Bahr, "Mechanical Behavior Assessment of Sucrose Using Nanoindentation," Jour. Mater. Res., **22**(7), 2037 (2007).

R. J. Roberts, R. C. Rowe, and P. York, "The Relationship Between Hardness and Cohesive Energy for Some Organic Crystals," Pharmeceutical Sdiences, **1**, 501 (1995).

J. S. Rowlinson, *Cohesion—A Scientific History of Intermolecular Forces*, Cambridge University Press, Cambridge, UK (2002).

A. Sasaki and M. Iwata, "Anisotropy of Microhardness in Anthracene Single Crystals," Phys. Stat. Sol. A, **85**, K105 (1984).

M. Schwoerer and H. C. Wolf, *Organic Molecular Solids*, Wiley-VCH Verlag GmbH, & Co. KGaA., Weinheim, Germany (2007).

J. Stephens, T. Gebre, A. K. Batra, M. D. Aggarwal, and R. B. Lal, "Microhardness Studies on Organic Crystals," Jour. Mater. Sci. Lett., **22**, 179 (2003).

L. Stryer, *Biochemistry—3rd Edition*, W. H. Freeman & Co., New York, USA (1988).

N. Vaidya, M. J. Joshi, B. S. Shah, and D. R. Joshi, "Knoop Hardness Studies on Anthracene Single Crystals," **20**, 333 (1997).

N. Vaidya and B. S. Shah, "Absolute Hardness of Phenanthrenedoped Anthracene Single Crystals by Knoop Indentation Technique," Indian J. Phys., **77A**, 59 (2003).

13   Polymers

13.1   INTRODUCTION

From a mechanical viewpoint, there are two broad classes of polymers: those that are thermoplastic (not cross-linked) and the thermosets (cross-linked). In the former case, molecular weight (length) is of considerable importance, but it is so large in the latter case that it no longer plays a role. Within the two broad classes atomic composition also varies. Thus, given the three parameters—molecular weight, amount of cross-linking, and atomic composition—the properties of polymers vary over a very large domain almost continuously. This makes a description of their behavior extremely unwieldy.

Few, if any, polymeric materials have exceptional physical properties. Nevertheless, by volume, they are the most widely used of any materials. This is because the raw materials used for making them are inexpensive, and especially because they are easy to convert into any desired shape.

Structural arrangements in polymers can be exceedingly complex. Crystals are rare, but not unknown. By first growing monomer crystals of diacetylene molecules, and then photo-polymerizing them, large optical-quality polydi-acety-lene crystals can be made, for example.

Because of their propensity for being disordered, polymeric solids are some-times called "organic glasses," but some polymers are inorganic, so this is not a good practice.

Disorder in a polymer specimen can be converted into order simply by stretching it, by extruding it, or by carefully orienting the long molecules during its solidiification. This strengthens it considerably in the dimension parallel to the unique axis. Similarly, strong membranes can be made by "blowing" polymeric tubes to cause stretching in two dimensions. In both cases the materials range from being somewhat to being very anisotropic.

Clearly, the hardnesses of thermoplastic polymers are not intrinsic. They depend on various extrinsic factors. Only trends can be cited. For example, as the molecular weight in polyethylene materials increases, they become harder. And, as the molecular aromaticity increases, a polymeric material becomes harder. Thus, higher molecular weight anthracene is harder than napthalene and more aromatic Kevlar is harder than polymethacrylate.

Chemistry and Physics of Mechanical Hardness, by John J. Gilman
Copyright © 2009 John Wiley & Sons, Inc.

There are, in general, two kinds of chemical bonding in polymers. Between individual molecules, the bonding consists of London's dispersion forces, but within the molecules there are much stronger covalent bonds. Cross-links are covalent, so they strongly bond pairs of molecules locally. Therefore, materials with high densities of cross-links have much higher melting points and hardnesses than do thermoplastics. Also, their properties tend to be isotropic.

The variety of polymers seems endless. First of all, they can have various "backbones." Not just carbon chains, but also chains of any elements in the block:

$$
\begin{array}{cccc}
B & C & N & O \\
 & Si & P & S \\
 & & As & Se \\
 & & & Te
\end{array}
$$

and mixtures of these elements along the lengths of chains. Second, the chains can have a multitude of lengths from very short to very long. Third, the chains can be cross-linked to form ribbons, tubes, membranes, and scaffolds. Fourth, residues of any imaginable size and shape can be attached on the sides of the chains. Fifth, metal atoms can be incorporated into the chains, as well as the residues.

Given the great variety outlined above, only some general trends for the hardnesses of polymers can be discussed here. For more detail regarding thermoplastic polymers, the reader is referred to Baltá, Calleja and Fakirov (2000).

13.2 THERMOSETTING RESINS (PHENOLIC AND EPOXIDE)

The first widely used synthetic polymer was phenol formaldehyde (Bakelite). It is made by heating phenol (C_6H_5OH—hydroxybenzene) together with formalde-hyde (H_2CO). These react to yield a three-dimensionally cross-linked polymer. To reduce the brittleness of Bakelite, it is usually filled with fibers or platelets of an inert solid. It is a good electrical insulator, relatively hard, and thermally stable to a few hundred degrees Centigrade. Its hardness is 50–60 kg/mm^2 (Mott, 1956).

Epoxy resins (di-phenolic chains) are closely related to phenol formaldehydes and are widely used to make reinforced composites with glass or carbon reinforcing fibers. Their monomers are cross-linked at lower temperatures than phenolic formaldehydes. Typical hardnesses for them are $H_v = 4.4$ kg/mm^2 (Olivier, et al., 2008).

The hardnesses of a few other resins are:

Cellulose nitrate	12 kg/mm^2
Vinyl resin	15 ″
Polystyrene	20 ″

These are approximate numbers because the sizes of indentations in them are time dependent, but the values do indicate that these materials are relatively soft.

A widely used polymeric resin for making construction laminates (Formica), low-cost dinnerware, and so on, is melamine ($C_3N_6H_6$) formaldehyde. It is harder than phenol formaldehyde.

13.3   THERMOPLASTIC POLYMERS

The first of the thermoplastic synthetic polymers to be developed was celluloid, made by combining nitrated cellulose (pure cotton subjected to nitric acid) and camphor ($C_{10}H_{16}O$), a plasticizer. The motivation was a search for a replacement for the ivory used in making billiard balls. It became a commercial product circa 1865, and is still used for making ping-pong balls.

Thermoplastics consist of long chains of monomer units. The bonding within the chains consists of strong covalent bonds, while the bonding between the chains consists of the relatively weaker Van der Waals (London) type. The resulting solids soften when heated to relatively low temperatures. They exhibit a variety of mechanical behaviors. Creep at relatively low temperatures is commonly observed for almost all of them, so their hardnesses are not quantitatively well-defined, although there are distinct qualitative differences. For example, it is quite clear that high density polyethylenes used for piping are harder than low density polyethylenes used for milk bottles. The former have VHNs $\approx 30\,kg/mm^2$, while the latter have VHNs $\approx 80\,kg/mm^2$.

In addition to the lengths of polymer molecules, the cross-sectional shapes have a major effect on their hardness and thermal stability. Aliphatics (paraffins, polyethylene, etc.) have the most simple cross-sectional shapes. Their simple and relatively symmetric shapes allow them to slide past on another readily via a process called *reptation* (de Gennes, 1990). As a result, linear polyethylene is relatively soft (Figure 13.1).

The cross-sectional shapes of Nylon chains have oxygen and nitrogen atoms protruding so they are less symmetric than polyethylene chains. Along with

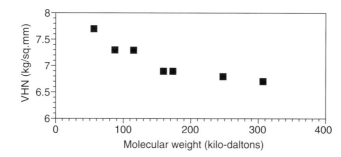

Figure 13.1 Dependence of the hardness of linear polyethylene on its molecular weight. Data from Baltá-Calleja et al., (1990).

other factors, this makes Nylon considerably harder than polyethylene, nearly ten times harder (VHN = 80–90 kg/mm^2).

Chains that include aromatic rings (phenols, pyridines, etc.) are said to be *polycyclic* and are stiffer, harder and more stable than aliphatic chains. Polycarbonate is an example, being hard enough for use in eyeglass lenses. An extreme example is Kevlar fiber.

Cross-linking of the chains produces substantial increases in hardness and thermal stability. However, it also reduces toughness, that is, increases brittleness.

A general characteristic of polymers is that their hardnesses tend to be proportional to their elastic moduli, particularly their shear moduli (Flores et al., 2000). However, the shear modulus is often anisotropic so an average value may not be an appropriate measure of hardness. The modulus for the plane of shear should be a better indicator.

13.4 MECHANISMS OF INELASTIC PLASTICITY

Inelastic deformation of any solid material is heterogeneous. That is, it *always* involves the propagation of localized (inhomogeneous) shear. The elements of this localized shear do not occur at random places but are correlated in a solid. This means that the shears are associated with lines rather than points. The lines may delineate linear shear (dislocation lines), or they may delineate rotational shear (disclination lines). The existence of correlation means that when shear occurs between a pair of atoms, the probability is high that an additional shear event will occur adjacent to the initial pair because stress concentrations will lie adjacent to it. This is not the case in a liquid where the two shear events are likely to be uncorrelated.

Correlated plastic deformation in polymers is very evident in the "necking" of polymeric rods or filaments. This is one form of inhomogeneous deformation. Experiments with Nylon filaments, for example, have shown that necks in them behave quite similarly to the Lueders bands observed in steel (Dey, 1967). This strongly suggests that plastic deformation in Nylon is associated with the motion of dislocations.

Disclinations can be expected when shear occurs between two parallel polymer chains (Gilman, 1973). This has been postulated to account for anelastic relaxation in some polymers at low temperatures. A general discussion of disclinations in polymers has been given by Li and Gilman (1970).

13.5 "NATURAL" POLYMERS (PLANTS)

The most common of all natural polymers is *cellulose*. It is ubiquitous in plant life in various molecular modifications and structural arrangements. Large quantities are found in the trunks, branches and leaves of trees as well as in

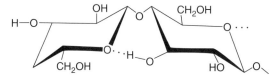

Figure 13.2 Schematic glucose dimer—the monomer of the cellulose polymer. The dimer consists of two 5-carbon, 1-oxygen rings linked by an oxygen bridge and a hydrogen bond so the rings lie in a plane, making the polymer resemble a ribbon. Two of the carbons in each ring have hydroxyl groups attached, two are connected to oxygen bridges and the fifth one has an aldehyde group attached. Adapted from Stryer (1975).

Table 13.1

Wood	Janka hardness (lbf)
Hickory	2400
White ash	1600
Yellow poplar	400
Red oak	1100
Sugar maple	1800
Redwood	410
White pine	430
Douglas fir	740
Southern pine	1200
Ponderosa pine	700

grasses and other plants. Cellulose is a polysaccharide chain in which diglucose monomers are linked by oxygen atoms (Figure 13.2). The chains are somewhat ribbon-like so they pack together well and are held laterally to their neighbors by hydrogen bonds. Thus they form solid rigid masses. In wood, cellulose is embedded in a matrix of *lignocellulose* (a mixture of glassy *hemicellulose* and *lignin*). The lignin fraction is large (up to 30 percent) and is a complex highly branched polyaromatic molecule with a very high molecular weight. It makes an important contribution to the strength of wood and the transport of water through it.

The hardness of wood varies markedly from soft balsa to hard ironwood with pine, oak, and maple in between. It is measured either by determining the force needed to push a hard ball (diameter = 0.444 in) into the wood to a depth equal to half the ball's diameter (Janka hardness); or by the initial slope of the force *vs.* penetration-depth curve (Hardness modulus). Average values of Janka hardnesses for typical woods are listed in Table 13.1. The data are from Green et al., (2006), and are for penetration transverse to the tree axis. The values are for moisture contents of about ten percent. The first set of five items are "hardwoods," while the second set are "softwoods." To roughly convert Janka hardnesses to VHN multiply by 0.0045.

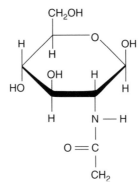

Figure 13.3   Monomer of chitin. Similar to the rings of the cellulose monomer except for the glucosamine group attached to one of the carbons.

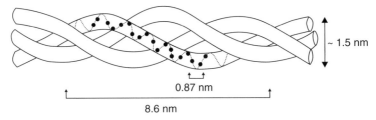

Figure 13.4   Collagen.

13.6   "NATURAL" POLYMERS (ANIMALS)

In the world of animals the most common polymer is *chitin*. It is closely related to cellulose. The main difference is that its monomers contain amino groups (Figure 13.3). Chitin is the material that forms the "shells" of scorpions, crabs, lobsters, and shrimp as well as the wings of insects. Chitin is not a particularly hard material, but it is relatively tough. It is a polysaccharide since its building blocks are glucose, but it contains some nitrogen. This distinguishes it from cellulose, which is purely hydrocarbon.

The most common structural polymer in *mammals* is *collagen* which is composed of three helical strands of poly-aminoacids (polypeptides) with a high concentration of glycine (30 percent). Each of the helical strands is itself composed of a helical polypeptide molecule (Figure 13.4). Furthermore, the collagen microfibrils twist together to form macrofibrils. The hierarchy of twistings contributes to the compliance and toughness of connective tissues in animals (e.g., muscle membranes and tendons).

REFERENCES

F. J. Balta'-Calleja, C. Santa Cruz, R. K. Bayer, and H. G. Kilian, "Microhardness and Surface Free Energy in Linear Polyethylene: The Role of Entanglements," Colloid & Polymer Sci., **268**, 440 (1990).

F. J. Baltá Calleja and S. Fakirov, *Microhardness of Polymers*, Cambridge University Press, Cambridge, UK (2000).

B. N. Dey, "Plastic-Flow Rates in Nylon Interpreted in Terms of Dislocation Motion," Jour. Appl. Phys., **38**, 4144 (1967).

A. Flores, F. J. Baltá Calleja, G. E. Attenburrow, and D. C. Bassett, "Microhardness Studies of Chain-extended PE: III. Correlation with Yield Stress and Elastic Modulus," Polymer, **41**, 5431 (2000).

P.-G. de Gennes, *Introduction to Polymer Dynamics*, Cambridge University Press, Cambridge, UK (1990).

J. J. Gilman, "Plastic Relaxation via Twist Disclination Motion in Polymers," Jour. Appl. Phys., **44**, 2233 (1973).

D. W. Green, M. Begel, and W. Nelson, "Janka Hardness Using Nonstandard Specimens," Research Note #FPL-RN-0303, Forest Products Laboratory, U.S Dept. of Agriculture, Madison, WI, USA (2006).

J. C. M. Li and J. J. Gilman, "Disclination Loops in Polymers," Jour. Appl. Phys., **41**, 4248 (1970).

B. W. Mott, *Micro-indentation Hardness Testing*, p. 247, Butterworths Scientific Publications, London, UK (1956).

L. Olivier, N. Q. Ho, J. C. Grandidier, and M. C. Lafarie-Frenot, "Characterization by Ultra-Micro Indentation of an Oxidized Epoxy Polymer," Pol. Degrad. & Stab., **93**, 489 (2008).

L. Stryer, *Biochemistry—3rd Ed.*, p. 261, W. H. Freeman & Co., New York, USA (1975).

14 Glasses

14.1 INTRODUCTION

Glasses are defined here to be supercooled liquids, that is, liquids that have been cooled so quickly that they have not had time to crystallize before atomic movements in them became very slow. The electronegative elements with affinity for two valence electrons (O, S, Se, and Te) can readily form chains of atoms and therefore relatively stable liquids which can be quenched to form glasses. Pure oxygen's mass is too low to do this at normal temperatures, but if it is combined with Si, P, Al, and so on to increase its effective mass, it is an outstanding glass former. Silica (SiO_2) does not readily crystallize to form quartz, so it readily takes the form of a glass. Numerous other covalently bonded compounds do the same, both simple and complex ones.

The definition of a glass used here differs from materials without any atomic order made, for example, by condensing a vapor. Glasses made by supercooling do have short-range order in them because the atomic structures of liquids are not random. Liquids usually have "association" in them, consisting of entities ranging from preferential pairing of atoms to long chains of atoms. Polymeric liquids are examples of the latter, but inorganic compounds also often form highly associated liquids. Examples are silicates, phosphates, and borates.

It is very difficult to cool pure metals and other pure elements fast enough to form glasses. However, metallic alloys can often be converted into glasses, particularly if they contain a mixture of small and large atoms such as iron and boron, or they are multi-component mixtures of metals that crystallize into more than one intermetallic compound (i.e., eutectic compositions). Thus, covalent chemical interactions of the atoms are important because they stabilize liquids and thereby inhibit crystallization.

High polymers do not readily crystallize because of their large lengths, so they are sometimes called "organic glasses." However, there are many inorganic polymers, so this is not accurate terminology. They have been discussed in the previous chapter. Figure 14.1 is a schematic comparison of dislocation lines in a crystalline and a glassy structure.

Chemistry and Physics of Mechanical Hardness, by John J. Gilman
Copyright © 2009 John Wiley & Sons, Inc.

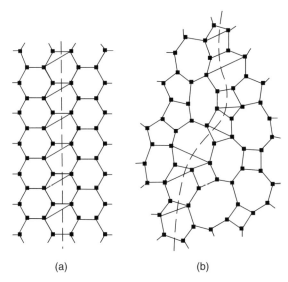

(a)   (b)

Figure 14.1 Schematic comparison of dislocation lines in a crystalline and a glassy structure. Dashed line indicates the center of a dislocation line. The vectors indicate the displacement of the atoms in the next level above the plane of the figure. At (a) the displacement (Burgers) vectors In the periodic crystal have a constant value. At (b) the displacements in the glass fluctuate in both magnitude and direction.

14.2   INORGANIC GLASSES

General discussion of the properties of inorganic glasses may be found in the book by Doremus (1994), for example.

The hardnesses of ordinary glasses are comparable with those of quenched and tempered steels, ranging from about VHN = 450 to 650 kg/mm². Among the hardest inorganic glasses is fused silica at about 700 kg/mm². This is considerably softer than quartz (VHN ≈ 1200 kg/mm²). The difference is partially accounted for by the fact that fused silica is about 20 percent less dense than quartz (SG = 2.21 vs. 2.65).

There are so many compositions of glass that generalizations regarding their hardnesses are difficult to make. Some trends can be found. For example, high alumina glasses tend to be harder than silica glasses; high glass-transition temperatures correlate with high hardness; and so on. The trends suggest that the hardnesses depend on bond strengths as in other materials, but various other factors affect the reported values. The other factors include the load on the indenter, the size of the indentation, the time duration of the indentation process, and the environment of the indenter (i.e., humidity, etc.). Li and Bradt (1992) showed that the difference observed for indenters immersed in water

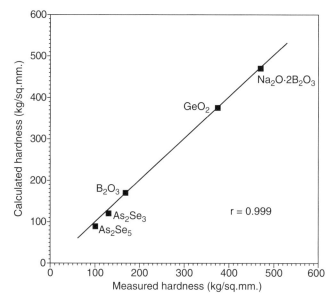

Figure 14.2 Comparison of calculated and measured hardnesses of non-silicate glasses (after Yamane and Mackenzie, 1974).

versus toluene is about 15 percent for the hardness of fused silica. These authors also tabulated data for fused silica that shows a spread from 378 to 1330 (Ave. = 690 kg/mm^2).

The susceptibility of hardness measurements of silica and silicate glasses to environmental factors is consistent with the effects of water on the deformation of quartz. The load effect and indentation size effect appear to be a result of the frictional forces at the indenter-specimen interfaces.

It is very difficult to obtain values for the intrinsic hardnesses of silicate and related types of glass. Therefore, no attempts at quantitative analyses will be made here. A semi-empirical method has been proposed by Yamane and Mackenzie (1974) based on the geometric mean of: bond strength relative to silica, shear modulus, and bulk modulus. For 50 silicate glasses it yields estimates within ten percent of measured values, and for a few non-silicate glasses it is quite successful, as Figure 14.2 indicates.

The success of the Yamane and Mackenzie method reinforces the idea that bond strengths are the key to hardness in glasses. This indicates that to make harder glasses, strong bonds need to be part of the composition. Figure 14.3 illustrates this for borosilicate optical glasses where the effect of the silica content on the hardness of the glass is shown when the remainder of the composition is B$_2$O$_3$ and K$_2$O in the ratio 5/3. This approach of adding strong bonds to increase hardness was also used by Makishima et al., (1983) to make a high

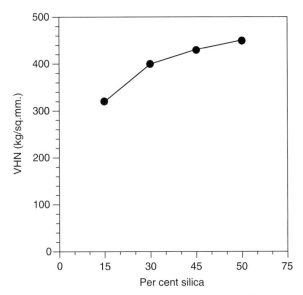

Figure 14.3   Effect of silica content on the hardness of borosilicate optical glass. The ratio of B_2O_3 to K_2O is 0.54 in these glasses. The data are from Schmidt and Reichardt (1986).

hardness glass $(1200 \, kg/mm^2)$ by increasing the nitrogen content to 18 percent.

When silica glass is indented at room temperature (low relative to the melting point) a large part of the deformation at the indentation is a result of densification rather plastic shearing (Neely and Mackenzie, 1968). This is analogous with the phase transformations that occur at indentations in covalent crystals (Chapter 5). In both cases, the initial structure is open since the initial coordination number of the atoms is only four. During densification at heterogeneous locations, it can, in principal proceed to become six, eight, twelve, and slightly beyond. Some of the densification may recover during unloading.

The hardnesses of several glasses have been shown by Prod'homme (1968) to correlate with their viscosities at room temperature (Figure 14.4). The viscosities were estimated by extrapolation from measured values at elevated temperatures because they were too high to measure at room temperature. The figure is a semi-logarithmic plot so the hardnesses increase slowly with viscosity. The rank-order is as expected with silica being the hardest, followed by the high boron glasses, then the ordinary crowns, and the soft Selinium at the bottom. No rationalization of the exponential dependence comes to mind.

A similar connection between hardness and glass viscosity is indicated in Figure 14.5 where hardness is plotted against softening temperature (the temperature at which the viscosity becomes 10^{11} Poise). Since the correlation coefficient is 0.92, the linear relationship is reasonably good.

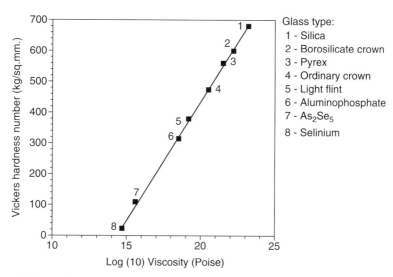

Figure 14.4   Relationship between hardnesses of various inorganic glasses and their viscosities. Data are from Prod'homme (1968).

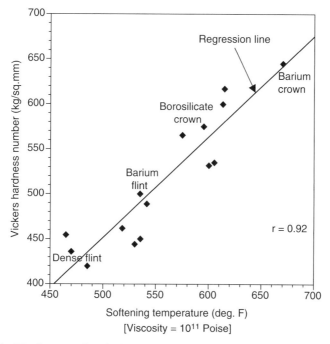

Figure 14.5   Hardnesses of optical glasses *versus* softening temperatures at which their viscosities become approximately 10^{11} Poise. A few of the points are labeled with their names. Data from Bastick (1950).

14.3 METALLIC GLASSES

In multi-component liquids, stabilization of the liquid is revealed by the formation of eutectics where the freezing temperature is suppressed. In such liquids, the atomic species (say A and B) are not distributed at random. There are more associated AB pairs (or other clusters) than expected for a random distribution. As a result in binary metal-metalloid alloys, such as Fe-B, the low melting-point eutectics occur at preferential compositions. The most common of these is at about 17 at. % B, or an atom ratio of one B for five Fe atoms (Gilman, 1978). This suggests that clusters of metal atoms surrounding metalloid atoms form (trigonal bipyramids). These probably share corners, edges, and faces.

Metal-metalloid alloys when cooled fast enough become produce glasses. The required cooling rates are of the order of $10^6 °C/sec$. These rates are achieved using cold copper quenching-wheels to make ribbons and sheets. At lower cooling-rates multi-component mixtures of several transition metals may form glasses. These are sometimes called "bulk glasses" (Greer and Ma, 2007), but this is a special definition of the word "bulk."

The hardnesses of metallic glasses vary with composition and temperature. The hardness is determined by the mobilities of dislocations in them (Gilman, 1968).

It has sometimes been argued that since glasses do not have periodic structures, plastic deformation in them is not mediated by dislocations. Then, since shear deformation in them cannot occur concertedly because not enough force is available, it is further argued that plastic deformation occurs through diffusive processes. As a matter of fact, however, flow is observed to occur much too rapidly in many of them for diffusion to be in control at temperatures below the glass-transition temperature. The rapidity of the flow is indicated by the fact that "tin cry" can readily be heard in several glasses while they are being deformed. The crying noise indicates the formation of glide bands within fractions of milliseconds (i.e., at nearly the speed of sound). This has been confirmed by systematic studies of acoustic emission from glasses being deformed (Vinogradov and Khonik, 2004).

On the surfaces of polished metallic glasses, very sharp, distinct offsets are observed at room temperature, indicating large and highly localized deformation that must be associated with the propagation of dislocations (Pampillo, 1972). The experimental observations have been reviewed by Li (1976).

Crystals have played a dominant role in the development of experimental knowledge about dislocations. Thus, it is often forgotten that the concept of dislocations was developed within the theory of continuous elastic solids. The very name was coined by A. E. H. Love, an elastician (Love, 1944). Therefore, dislocations need not have fixed displacement vectors.

In glasses, dislocation lines are the boundaries between plastically sheared areas, and material in which plastic shear has not yet occurred (Gilman, 1968).

These lines do not have displacement vectors with single valued magnitudes or directions, as are found in crystals. Their displacement vectors (Burgers vectors) fluctuate with magnitudes that are narrowly distributed around the "mesh size" of the structure. This is because the elastic energy depends on the square of the displacements across dislocation lines. Otherwise the force (i.e., the work) needed to cause the shear becomes very large (Chapter 4). When shear occurs locally, the probability that a subsequent shear will occur adjacent to it is much larger than that it will occur at some unrelated place. Thus correlated plastic shear activity tends to occur at dislocations in both periodic (crystal) structures, and in nonperiodic structures (glasses, granular media, etc.). In the latter case, suppose the characteristic size of the microstructure is Δ (the shear displace-ments), and the fluctuations in the shear displacements are $\pm\delta$. Then, since the elastic part of energy of a dislocation is proportional to the square of the displacement, $(\Delta \pm \delta)^2$, the ratio of the elastic energy with the fluctuations to the energy without fluctuations is $[1 + (\delta/\Delta)^2]$. Thus, fluctuations as large as 30 percent cause only a 10 percent increase in the elastic energy.

The average value of Δ must be conserved over long distances to minimize both the elastic energy and the chemical (core) energy. Also, there will be little tendency for a dislocation line to remain in a single plane. It will tend to follow the plane of maximum shear stress. This is observed experimentally.

14.3.1 Hardness—Shear Modulus Relationship

The stress needed to move a dislocation line in a glassy medium is expected to be the amount needed to overcome the maximum barrier to the motion less a stress concentration factor that depends on the shape of the line. The macro-scopic behavior suggests that this factor is not large, so it will be assumed to be unity. The barrier is quasi-periodic where the quasi-period is the average "mesh" size, Δ of the glassy structure. The resistive stress, initially zero, rises with displacement to a maximum and then declines to zero. Since this happens at a dislocation line, the maximum lies at about $\Delta/4$. The initial rise can be described by means of a shear modulus, G, which starts at its maximum value, G_0, and then declines to zero at $\Delta/4$. A simple function that describes this is, $G = G_0 \cos (4\pi x/\Delta)$ where x is the displacement of the dislocation line. The resistive force is then approximately $G(x) \Delta^2$, and the resistive energy, U, is:

$$U = G_0\Delta^2 \int_0^{\Delta/4} \cos(4\pi x/\Delta)\,dx = G_0\Delta^3/4\pi \qquad (14.1)$$

The work, W, done by the applied shear stress, τ, is approximately $W = \tau\Delta^3$. Equating U and W yields $\tau = G_0/4\pi$. Then, since $\tau = Y/2$, the yield stress Y is given by $Y \approx G_0/2\pi$.

Metallic glasses are almost elastic-perfectly plastic, so indentations in them are limited by the critical shear stress, not by strain-hardening as in crystalline

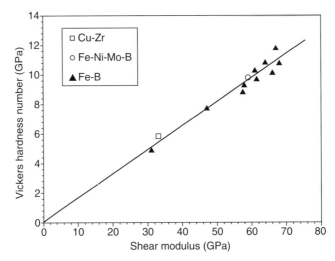

Figure 14.6   Hardnesses of metallic glasses vs. their elastic shear moduli. Data from Davis et al., 1994. The glass compositions are: $Cu_{68}Zr_{32}$, $Fe_{40}Ni_{38}Mo_4B_{18}$, FeB (various). The line in the graph has a slope of $G/2\pi$.

metals. Therefore, their hardness numbers are expected to be $H \approx G_0/2\pi$. This is just what is observed (Figure 14.5). The figure shows data for various glasses (mostly metal-metalloid ones) as a function of their shear moduli (Davis et al., 1994).

For a variety of transition metal glasses, the hardness-shear modulus correlation is shown in Figure 14.6 (Johnson and Samwer, 2005). Here the hardness data were derived from "yield stresses" using the ratio of about three that has been observed by Stoica et al. (2005). The variation is linear and it agrees quantitatively with the simple model that led to Equation 14.1. The average value of H/G is $1/2\pi$ for the data of Figure 14.6. Thus for both metal-metalloid and transition-metal glasses, the Chin-Gilman parameter averages 0.16. This suggests that the chemical bond type in both cases is covalent. This further suggests that glass-forming liquids are stabilized by short-range covalency between hetero-atom pairs (Figure 14.7).

Figure 14.8 shows how the hardness varies with the boron content of FeB glasses. The variation is nearly linear over the glass forming range.

The highest hardness, $1250\,kg/mm^2$ ($12.5\,GPa$) in Figure 14.5, corresponds to a yield stress of about $600,000\,psi$. which is similar to that of heavily cold-drawn steel. This hardness is about that expected from the strength of FeB chemical bonds. The enthalpy (heat) of formation, ΔH_f, of FeB is about $70\,kJ/mol. = 0.73\,eV$. while the molecular volume is about $9.6\,\text{Å}^3$ so the atomic shear strength is expected to be about $\Delta H_f/2V_m = 6\,GPa$. Then the expected hardness is about $1800\,kg/mm^2$, which is not far from the observed value. Thus, the barriers to dislocation motion appear to be FeB bonds. Note: this hardness is about six times larger than that of the hardest glasses composed only of

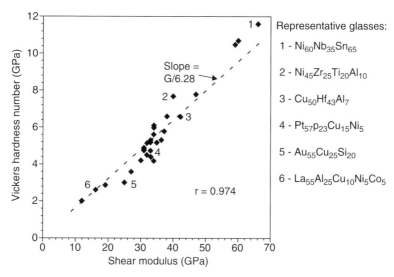

Figure 14.7   Hardesses of transition metal glasses *vs.* their shear moduli. Representative compositions are indicated by numbers corresponding to the list on the right in the figure.

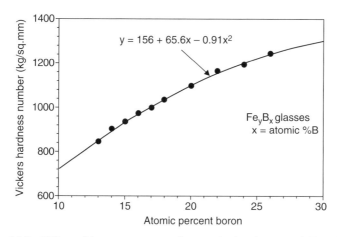

Figure 14.8   Effect of boron concentration on the hardnesses of *FeB* glasses.

transition metals (no metalloid atoms). This ratio is similar to the difference in the bond strengths between metal-metalloid and metal-metal interactions.

Two outstanding properties of FeB metallic glasses are their low magnetic permeabilities and their low acoustic attenuations. The former results from their lack of magnetic anisotropy and has led to their use in power transformers, theft detectors, and various electronic devices. The latter results from the very low dislocation mobility in them.

14.3.2   Stable Compositions

Metallic glass compositions based on Fe and Ni are thermally stable only up to a few hundred degrees where they begin to flow readily and devitrify. However, they can be based on more refractory metals (e.g., $Mo_{60}Fe_{20}B_{20}$), and are then stable to temperatures above $900\,°C$. Very high hardness can also be achieved, for example, $Mo_{40}Fe_{40}B_{20}$ glass has a hardness of $1950\,kg/mm^2$ (Ray and Tanner, 1980).

Also, by mixing several transition metals glass-forming liquids can be stabilized so relatively slow cooling-rates will form glasses. This allows thicker cross-sections to be obtained with glassy structures. Such glasses have come to be known as "bulk" metallic glasses.

REFERENCES

R. E. Bastick, in discussion section of paper: E. W. Taylor, "The Plastic Deformation of Optical Glass," Jour. Soc. Glass Tech., **34**, 75 (1950).

L. A. Davis, S. K. Das, J. C. M. Li, and M. S. Zedalis, "Mechanical Properties of Rapidly Solidified Amorphous and Microcrystalline Materials: A Review," Inter. Jour. Rapid Solidif. **8**, 73 (1994).

R. H. Doremus, *Glass Science—2nd Edition*, J. Wiley & Sons, New York, USA (1994).

J. J. Gilman, "The Plastic Response of Solids," in *Dislocation Dynamics*, Edited by A. R. Rosenfield, G. T. Hahn, A. L. Bement, and R. I. Jaffee, McGraw-Hill Book Company, New York, USA, p. 16 (1968).

J. J. Gilman, "Structure of Ferrous Eutectic Liquids," Phil. Mag., **37**, 577 (1978).

A. L. Greer and E. Ma, "Bulk Metallic Glasses: At the Cutting Edge of Metals Research," MRS Bull., **32**, 611 (2007).

W. L. Johnson and K. Samwer, "A Universal Criterion for Plastic Yielding pf Metallic Glasses with a $(T/T_g)^{2/3}$ Temperature Dependence," Phys. Rev. Lett., **95**, 195501 (2005).

H. Li and R. C. Bradt, "The Indentation Load/Size Effect and the Measurement of the Hardness of Vitreous Silica," Jour. Non-Cryst. Sol. **146**, 197 (1992).

J. C. M. Li, "Defect Mechanisms in the Deformation of Amorphous Materials," *Chapter 16 in Frontiers in Materials Science*, Edited by L. E. Murr and C. Stein, Marcel Dekker Inc., New York (1976).

A. E. H. Love, *Elasticity*, p. 121, Dover Publications, New York, USA (1944).

M. Makashima, N. It. Mitomo, and M. Tsutsumi, "Microhardness and Transparency of an La-Si-O-N Oxynitride Glass," Comm. Amer. Cer. Soc., C55, March 1983.

J. E. Neely and J. D. Mackenzie, "Hardness and Low-Temperature Deformation of Silica Glass," Jour. Mater. Sci., **3**, 603 (1968).

C. A. Pampillo, "Localized Shear Deformation in a Glassy Metal," Scripta Met., **6**, 915 (1972).

M. Prod'homme, "Some Results Concerning the Microhardness of Glasses," Phys. & Chem. Glasses, **9** (3), 101 (1968).

R. Ray and L. E. Tanner, "Molybdenum-based Metallic Glasses," Jour. Meter. Sci., **15**, 1596 (1980).

G. Schmidt and H. Reichardt, "Mechanische Eigenschaften von Versuchs-schmelzen Optisher Glaeser," Silikattechnik, **37**, 183 (1986).

M. Stoica, J. Eckert, S. Roth, Z. F. Zhang, L. Schultz, and W. H. Wang, "Mechanical Behavior of $Fe_{65.5}Cr_4Mo_4Ga_4P_{12}C_5B_{5.5}$ Bulk Metallic Glass," Intermetallics, **13**, 764 (2005).

M. Yamane and J. D. Mackenzie, "Vicker's Hardness of Glass," Jour. Non-Cryst. Sol., **15**, 153 (1974).

A. Yu. Vinogradov and V. A. Khonik, "Kinetics of Shear Banding in Bulk Metallic Glasses Monitored by Acoustic Emission Measurments," Phil. Mag., **84**, 2147 (2004).

15 Hot Hardness

15.1 INTRODUCTION

There is a very large literature on the effect of temperature on hardness. A summary of it is that when materials get hot, they soften. However, in the opinion of this author, there is no satisfactory physical theory of this softening. It is commonly discussed in terms of the thermal activation theory. However, the latter applies to gas-phase chemical reactions (and less well to simple liquid-phase chemical reactions); that is, to reactions and interactions between discrete molecules. But the dislocation lines that determine hardness are not particles. Adjacent segments of the lines interact strongly. Also, the observed temperature dependence of hardness does not generally behave as if it is associated with well-defined (i.e., specific) activation energies. Finally, dislocations move at very low temperatures close to $0\,°K$ where the thermal energy density is negligible compared with the strain energy density induced by an applied stress.

A major difficulty with the thermal activation theory is that the applied stress is usually large during plastic indentation at low to moderate temperatures. Therefore, the internal energy per atomic volume associated with the applied stress is large, indeed very large, compared with the thermal energy per atom. This means that the thermal vibrations act only as a perturbation of the effects of the applied stress. The gradients of the stress-induced chemical potential are the primary driving forces. Plastic deformation occurs even if the temperature is close to $T = 0\,°K$.

Since there is no good physical framework in which the measured hardness versus temperature data can be discussed, descriptions of it are mostly empirical in the opinion of the present author. Partial exceptions are the elemental semiconductors (Sn, Ge, Si, SIC, and C). At temperatures above their Debye temperatures, they soften and the behavior can be described, in part, in terms of thermal activation. The reason is that the chemical bonding is atomically localized in these cases so that localized kinks form along dislocation lines. These kinks are quasi-particles and are affected by local atomic vibrations.

Hardness-temperature graphs are quite variable from one material to another. There are no standard shapes of the curves even when the

Chemistry and Physics of Mechanical Hardness, by John J. Gilman
Copyright © 2009 John Wiley & Sons, Inc.

temperatures are normalized by using ratios of the temperature values to melting points or to Debye temperatures (West brook, 1957).

Because the subject is largely empirical, no attempt at a systematic discussion of it will be made here. Nevertheless, hot hardness is of great technological importance so a few selected topics will be discussed. Hot hardness limits the performance of many engineering devices, particularly heat engines. Among the latter are steam engines, internal combustion engines, and gas turbines, as well as rockets. The efficiencies of these depends on the maximum operating temperature which is determined by on high temperature strength (hardness).

In almost all applications, hot hardness alone is not enough to define useful materials. Corrosion resistance, particularly oxidation resistance, is also needed. The surfaces of all metals react with oxygen. Some form protective oxide coatings, such as aluminum oxide and chromium oxide (stainless steel), but most do not. Therefore only a few pure metals are useful for making high temperature structures. Alloying is sometimes effective in markedly improving oxidation resistance. Familiar examples are; stainless steel in which chromium additions to iron result in the formation of protective chromate surface coatings. Another case, important for gas turbines, is the addition of aluminum to nickel to make nickel-based superalloys. The latter have the best known combination of hot hardness and hot oxidation resistance of all metallic alloys.

15.2   NICKEL ALUMINIDE VERSUS OXIDES

Figure 15.1 compares the hot hardnesses of the intermetallic compound Ni_3Al, which strengthens nickel-based alloys, and three oxides. It is apparent that the intermetallic compound, although hard, is not nearly as hard as the oxides which are ten times as hard at room temperature. Also, the oxides retain their hardnesses at high temperatures. At $800\,°C$, the garnet (YAG) remains seven to eight times as hard as the Ni_3Al and remains quite hard up to $1400\,°C$, which is nearly the melting point of steel ($1500\,°C$). At $1200\,°C$, yttrium aluminum garnet is the hardest of all known oxides.

Ni_3Al is one of a very few intermetallic compounds that is ductile at room temperature. Additionally, it has the interesting property that between 25 and $550\,°C$ it gets harder rather than softer. This "anomalous yielding" is not unique to this compound. Several other intermetallic compounds behave similarly.

Additional discussion of Ni_3Al is given in Chapter 8.

15.3   OTHER HARD COMPOUNDS

There are a number of intermetallic compounds that are hard at high temperatures (Fleischer and McKee, 1993). A few examples are $AlNb_2$, $AlZr_2$,

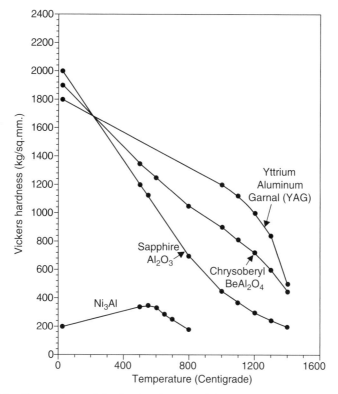

Figure 15.1   Comparison of the hot hardnesses of three strong oxides and the strong intermetallic compound, Ni_3Al. Yttrium aluminum garnet is the hardest of all oxides at high temperatures.

$Ir_3Ni_2Nb_5$, $Be_{17}Nb_2$, and SiV_3. However, they are not oxidation resistant, so they have limited usefulness. One that is harder than Ni_3Al at 1150 °C and has good ductility and oxidation resistance is AlRu (and AlRu with B).

Also metal-metalloid compounds tend to retain their hardnesses as temper-atures become elevated. These compounds have been discussed in Chapter 10.

15.4   METALS

Figure 15.2 shows some typical hardness data for a typical metal (copper) as a function of temperature. It indicates that there are usually two regimes: one above about half the melting temperature and one below. Both tend to be exponential declines, so they are linear on semi-logrithmic graphs. The temperature at which the "break" occurs is not strictly fixed, but varies from one metal to another, with the purity of a metal, with grain size, and so on.

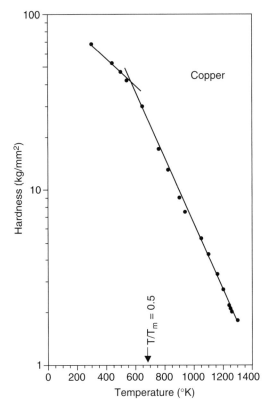

Figure 15.2   Log H plotted versus temperature for copper; a typical metal. The graph indicates the exponential decline of the hardness with increasing temperature, and the change in behavior at about half the melting point, T_m.

The two lines on graphs such as Figure 15.2 may each be described by equations of the form:

$$H = A \exp(-B/T) \tag{15.1}$$

where A and B are constants. Values of these constants have been determined for many metals by Westbrook (1953). When T becomes zero, Equation 15.1 indicates that $H = A$ so this constant is simply the limiting hardness at very low temperatures. It correlates roughly with the average shear moduli of the metals as might be expected. The slopes of the log H versus T lines yield values of the B constant. Westbrook found that these lie along a correlation curve when plotted against the reciprocals of the melting temperatures. Thus the slopes tend to be smaller the higher the melting temperature, but no simple interpretation of this has been given.

15.5   INTERMETALLIC COMPOUNDS

For compounds the hardness-temperature curves are similar to those for the pure metals. Semi-logarithmic graphs of the data show two straight lines with the "knees" at about half the melting temperatures. For a dozen aluminides, Petty (1960) shows this.

REFERENCES

R. L. Fleischer and D. W. McKee, "Mechanical and Oxidation Properties of AlRu-Based High-Temperature Alloys," Metall. Trans. A, **24A**, 759 (1993).

E. R. Petty, "Hot Hardness and Other Properties of Some Binary Intermetallic Compounds of Aluminum," Jour. Inst. Metals, **89**, 343 (1960–61).

J. H. Westbrook, "Temperature Dependence of the Hardness of Pure Metals," Trans. Amer. Soc. Met., **45**, 221 (1953).

J. H. Westbrook, "Temperature Dependence of the Hardness of Secondary Phases Common in Turbine Bucket Alloys," Trans. TMS-AIME, **209**, 898 (1957).

16 Chemical Hardness

16.1 INTRODUCTION

The concept of "chemical hardness" was originally developed as a measure of the stability of molecules. Its relationship to physical hardness and to solids is discussed here. Also, it is pointed out that shear moduli and polarizabilitites, as well as band gaps in covalent crystals, are related to it.

Chemical hardness is an energy parameter that measures the stabilities of molecules—atoms (Pearson, 1997). This is fine for measuring molecular stability, but energy alone is inadequate for solids because they have two types of stability: size and shape. The elastic bulk modulus measures the size stability, while the elastic shear modulus measures the shape stability. The less symmetric solids require the full set of elastic tensor coefficients to describe their stabilities. Therefore, solid structures of high symmetry require at least two parameters to describe their stability.

The formal definition of the electronic chemical hardness is that it is the derivative of the electronic chemical potential (i.e., the internal energy) with respect to the number of valence electrons (Atkins, 1991). The electronic chemical potential itself is the change in total energy of a molecule with a change of the number of valence electrons. Since the elastic moduli depend on valence electron densities, it might be expected that they would also depend on chemical hardness densities (energy/volume). This is indeed the case.

Physical hardness can be defined to be proportional, and sometimes equal, to the chemical hardness (Parr and Yang, 1989). The relationship between the two types of hardness depends on the type of chemical bonding. For simple metals, where the bonding is nonlocal, the bulk modulus is proportional to the chemical hardness density. The same is true for non-local ionic bonding. However, for covalent crystals, where the bonding is local, the bulk moduli may be less appropriate measures of stability than the octahedral shear moduli. In this case, it is also found that the indentation hardness—and therefore the Mohs scratch hardness—are monotonic functions of the chemical hardness *density*.

One implication of these findings is that chemical hardness is related to the band gaps of covalent crystals, consistent with its being related to the LUMO-HOMO gaps of molecules. Data indicate that this is indeed the case. Another implication is that the chemical potential of a solid is, in general, a second order tensor that is related to the deformation tensors (strains) through the fourth order tensor of elastic coefficients.

The theory of chemical and physical hardnesses is useful because it unifies understanding of various properties and it connects the behaviors of molecules with those of liquids and solids.

Stability is sometimes associated with the bulk modulus alone, but this is not valid because the bulk modulus of a liquid, and its corresponding solid, are nearly equal at the melting temperature, while their mechanical stabilities are very different. For example, take the case of aluminum. The bulk modulus of its liquid is about 0.3 Mbar, while that of its solid is about 0.7 Mbar, both measured near its melting point. On the other hand, the shear modulus of liquid aluminum is zero, while it is about 0.25 Mbar for solid aluminum.

16.2   DEFINITION OF CHEMICAL HARDNESS

Chemical potential, μ, is another name for total internal energy. Convenient units for it are energy per mole. In terms of the work done on (PdV), and the entropy (S) of a gaseous or liquid substance, it may be written in differential form (Callen, 1960):

$$d\mu = PdV - TdS \qquad (16.1)$$

where: P = pressure;   V = molar   volume = (v/N);   N = number   of   moles; T = temperature; and S = molar entropy. When T = 0, ignoring the zero-point energy, this reduces to:

$$d\mu = PdV \qquad (16.2)$$

For a solid this is more complex; P becomes the stress tensor, σ_{ij}, so the work done is $\sigma_{ij} \, dV$. But, $\sigma_{ij} = \Sigma_{kl} C_{ijkl} \varepsilon_{kl}$, so:

$$d\mu = -1/2 \, \Sigma_{ijkl} C_{ijkl} \varepsilon_{ij} \varepsilon_{kl} dV \qquad (16.3)$$

In general, C_{ijkl} is a 9×9 tensor with 81 terms, but symmetry reduces this considerably. Thus, for the cubic crystal system, it has only three terms (C_{1111}, C_{1212}, and C_{4444}) and for an isotropic material only two terms remain: B = bulk modulus and G = shear modulus. A further simplification is that the bulk modulus, B for the cubic system is given by ($C_{1111} + 2C_{1212}$)/3, and the two shear moduli are C_{44} and ($C_{1111} - C_{1212}$)/2.

Experimental data as well as density functional theory show that the ground-state properties of solids depend primarily on the densities of the valence electrons. Therefore, μ_E may be considered to be the *electronic chemical potential* (Pearson, 1997). Since μ_E denotes the energy per mole of electrons, and N is the number of electrons, if a change, dN, occurs while the number of positive nuclei, β remains constant, the energy change is:

$$dU = \mu_E + (\rho d\beta)$$ (16.4)

Therefore, if β doesn't change, we have Figure 16.1:

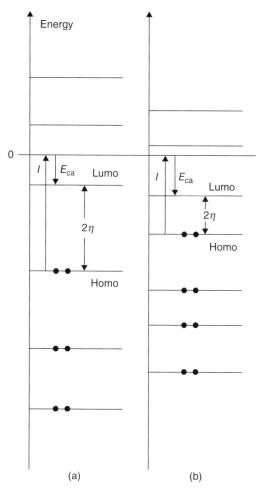

(a)   (b)

Figure 16.1   The chemical hardness of an atom, molecule, or ion is defined to be half. The value of the energy gap between the bonding orbitals (HOMO—highest orbitals occupied by electrons), and the anti-bonding orbitals (LUMO—lowest orbitals unoccupied by electrons). The "zero" level is the vacumn level, so I is the ionization energy, and A is the electron affinity. (a) For "hard" molecules the gap is large; (b) it is small for "soft" molecules. The solid circles represent valence electrons. Adapted from Atkins (1991).

$$\mu_E = (\partial U/\partial N)_\beta \tag{16.5}$$

In standard thermodynamics, N would be the number of molecules, and the chemical potential might be designated, μ_T, which is the negative of the Milliken electronegativity, X_M. This latter is the rate of change of the energy with changes in the number of electrons and is related to the ionization energy, I, and the electron affinity, A. Thus:

$$\mu_E = -X_M = (\partial U/\partial N)_\beta \approx -(I+A)/2 \tag{16.6}$$

That is, the Milliken electronegativity equals the slope of the U versus N "curve." The apos-trophes refer to the fact that it is a pseudo-curve since it is only defined for integral num-bers of electrons.

The electronic chemical hardness, η_E is the curvature of the U versus N "curve." Thus it is the second derivative of U with respect to N:

$$\eta_E = 1/2\left(\partial^2 U/\partial N^2\right)_\beta = (\partial\mu/\partial N)_{\beta,T} \tag{16.7}$$

Chemical hardness measures the resistance to a chemical change. This may be seen by considering a reaction between two molecules C and D in which electrons are transferred from D to C. Initially the two chemical potentials are:

$$\mu_C = X_C + 2\eta_C\Delta N$$
$$\mu_D = X_D - 2\eta_D\Delta N$$

After the reaction has occurred the chemical potentials must become equal. Solving for ΔN:

$$\Delta N = (\chi_C - \chi_D)/2(\eta_C + \eta_D) \tag{16.8}$$

Hence, the electronegativity difference drives the reaction, while the sum of the hardnesses resists the reaction.

16.3   PHYSICAL (MECHANICAL) HARDNESS

Yang, Parr, and Uytterhoeven (1987) have shown that chemical and mechani-cal hardnesses (physical stabilities) are connected. Consider the isotropic case, and differentiate Equation 16.5 with respect to N:

$$(\partial\mu/\partial N) = (-V/N^2)(\partial P/\partial N) \tag{16.9}$$

but the definition of the bulk modulus is:

$$\partial P = -B(\partial V/V)_{\alpha,T} \qquad (16.10)$$

differentiating this with respect to N:

$$\partial P/\partial N = -B/V(\partial V/\partial N)_{\alpha,T} \qquad (16.11)$$

setting N = one mole:

$$\eta_P = BV_m \qquad (16.12)$$

So, the physical hardness density, $\eta_{P/}V_m = B$. (In Parr's notation, the indentation hardness, H $(kg/mm^2) = (M/\rho)B$, where (ρ/M) is the number density of the atoms, but this does not agree very well with measured values, most solids being anisotropic.)

16.4   HARDNESS AND ELECTRONIC STABILITY

When atoms come together to form molecules, their orbitals combine in symmetric and antisymmetric pairs to form two principal sets of orbitals: those that bond the atoms and those that anti-bond. Pairs of electrons of opposite spin occupy the bonding orbitals which have a range of energies. The top of the range is called the HOMO level—highest occupied molecular orbital. The antibonding orbitals are of higher energy and also lie in a range, the bottom of which is called the LUMO level—lowest unoccupied molecular orbital. The two ranges are separated by an energy gap. Any electrons with enough energy to lie above the LUMO level weaken the bonding by exerting a pressure called the Schrödinger pressure. Thus the stability of a molecule is related to the size of the LUMO-HOMO gap.

To get an approximate expression for the chemical hardness, start with an expression for the electronic chemical potential. Let a hypothetical atom have an energy, U_o. Subtract one electron from it. This costs I = ionization energy. Alternatively, add one electron to it. This yields A = electron affinity. The derivative = electronic chemical potential = $\mu = \Delta U/\Delta N = (I + A)/2$. The hardness is the derivative of the chemical potential = $\eta = \Delta\mu/\Delta N = (I - A)/2$.

One more step provides an operational definition. The HOMO level lies, I = ionization energy, below the vacuum level, while the LUMO level lies, A = electron affinity, below it. Thus, the chemical hardness lies midway in the gap and usually is given in units of eV.

The bonding in solids is similar to that in molecules except that the gap in the bonding energy spectrum is the minimum energy band gap. By analogy with molecules, the chemical hardness for covalent solids equals half the band gap. For metals there is no gap, but in the special case of the alkali metals, the electron affinity is very small, so the hardness is half the ionization energy.

16.5 CHEMICAL AND ELASTIC HARDNESS (STIFFNESS)

Equation 16.12 expresses a relation between η and B. This is not a universal relation, but it does apply to the sp-bonded elements of the first four columns of the Periodic Table. Using chemical hardness values given by Parr and Yang (1989), and atomic volumes from Kittel (1996), it has been shown that the bulk moduli of the Group I, II, III, and IV elements are proportional to the chemical hardness density (CH/atomic volume) (Gilman, 1997). The correlation lines pass nearly through the coordinate origins with correlation coefficients, r = 0.999. Thus physical hardness is proportional to chemical hardness (Pearson, 2004).

The Group IV elements also show a linear correlation of their octahedral shear moduli, $C_{44}(111)$ with chemical hardness density $(E_g/2V_m)$. This modulus is for for shear strains on the (111) planes. It is a measure of the shear stiffnesses of the covalent bonds. The (111) planes lie normal to the bonds that connect the atoms in the diamond (or zinc blende) structure. In terms of the three standard moduli for cubic symmetry $(C_{11}, C_{12},$ and $C_{44})$, the octahedral shear modulus is given by $C_{44}(111) = 3C_{44}/1 + [4C_{44}/(C_{11} - C_{12})]$. Since the (111) planes have three-fold symmetry, they have only one shear modulus. The bonds across the octahedral planes have high resistance to shear which probably results from electron correlation in the bonds (Gilman, 2002).

16.6 BAND GAP DENSITY AND POLARIZABILITY

Since chemical hardness is related to the gaps in the bonding energy spectra of covalent molecules and solids, the band gap density (E_g/V_m) may be substituted for it. When the shear moduli of the III–V compound crystals (isoelectronic with the Group IV elements) are plotted versus the gap density there is again a simple linear correlation.

Another property that is related to chemical hardness is polarizability (Pearson, 1997). Polarizability, α, has the dimensions of volume polarizability (Brinck, Murray, and Politzer, 1993). It requires that an electron be excited from the valence to the conduction band (i.e., across the band gap) in order to change the symmetry of the wave function(s) from spherical to uniaxial. An approximate expression for the polarizability is $\alpha = \beta \ (N/\Delta^2)$ where β is a constant, N is the number of participating electrons, and Δ is the excitation gap (Atkins, 1983). The constant, $\beta = (qh)/(2\pi^2 m)$ with q = electron charge, m = electron mass, and h = Planck's constant. Then, if N = 1, $(1/\alpha)$ is proportional to Δ^2, and elastic shear stiffness is proportional to $(1/\alpha)$.

It is simple to understand the connection between the shear modulus and α. A sphere can be deformed into a prolate ellipsoid either by mechanical stress, or by an electric field. The input work required is measured by G = shear modulus in the first case and by α in the second case. Equating the input work needed in each case and solving for G, yields:

$$G = (3/4\pi)(q^2/r\alpha) \tag{16.13}$$

where r = atomic (molecular) radius (Gilman, 1997). For metals, α is the atomic polarizability. For non-metals, α is derived from the dielectric constant, $\varepsilon(0)$, using the Clausius-Mossotti equation. Polarizability underlies much mechanical behavior, and depends only on size at constant valence electron number.

16.7   COMPRESSION INDUCED STRUCTURE CHANGES

There have been many studies of crystal structure changes in high pressure cells, and these are referred to as "pressure-induced" changes. However, this terminology is unfortunate since true hydrostatic pressure is rarely achieved experimentally. The cells have fixed symmetry (shapes), but structure transformations require symmetry changes, so it is not possible to maintain hydrostatic conditions during a phase transition when the symmetry of the specimen changes while that of the pressure cell does not. Also, it has been shown for covalently bonded crystals, that most of the deformation during compression-induced phase changes results from bond-bending rather than bond-shortening (Gilman, 1993). Thus the values reported for the critical "pressures" of structure changes are unlikely to be truly pertinent.

In relatively recent years, it has been found that that indentations made in covalent crystals at temperatures below their Debye temperatures often result from crystal structure changes, as well as from plastic deformation via dislocation activity. Thus, indentation hardness numbers may provide better critical parameters for structural stability than "pressure cell" studies because indentation involves a combination of shear and hydrostatic compression and a phase transformation involves both of these quantities.

The connection between hardness and a measure of the ease with which a crystal can change its shape is its reciprocal polarizability (as shown above). The softer the crystal, the greater its polarizability. The hardnesses have the dimensions of pressure, and the polarizabilities are derived from the dielectric constants through the Clausius-Mosottii equation. That is:

$$\alpha_{cm} = (3/4\pi)(1/N)[(\varepsilon - 1)/(\varepsilon + 2)] \qquad (16.14)$$

where ε = dielectric constant, and N = number density of atoms (cgs units). Since $1/\alpha$ is proportional to the energy band-gap, it is also proportional to the chemical hardness.

An implication of the connection with polarizability is that structural transformations may occur when the external work done is just enough to equal E_g ($\Delta V/V$). That is, just enough to close the band-gap. This suggestion was originally made by Jamieson (1963), and was supported by studies of Gilman (1993).

An indication that hardness numbers are good indicators of structural stability is that they correlate quite well with critical transformation pressure values. This was shown in Gilman (2007) where theoretical values of critical pressures from Van Vechten (1973) are plotted along with VHN's for most of the III–V compounds. Theoretical values are used because they are internally more consistent than experimental values which depend on various conditions. Although some of the hardness numbers in the figure may not be accurate, the trend is very clear, and the correlation coefficient is high ($r = 0.997$).

16.8   SUMMARY

It is shown that the stabilities of solids can be related to Parr's "physical hardness" parameter for solids, and that this is proportional to Pearson's "chemical hardness" parameter for molecules. For sp-bonded metals, the bulk moduli correlate with the chemical hardness density (CHD), and for covalently bonded crystals, the octahedral shear moduli correlate with CHD. By analogy with molecules, the chemical hardness is related to the gap in the spectrum of bonding energies. This is verified for the Group IV elements and the isoelectronic III–V compounds. Since polarization requires excitation of the valence electrons, polarizability is related to band-gaps, and thence to chemical hardness and elastic moduli. Another measure of stability is indentation hardness, and it is shown that this correlates linearly with reciprocal polarizability. Finally, it is shown that theoretical values of critical transformation pressures correlate linearly with indentation hardness numbers, so the latter are a good measure of phase stability.

REFERENCES

P. W. Atkins, *Molecular Quantum Mechanics*, p. 384, Oxford University Press, New York, USA (1983).

P. W. Atkins, *Quanta*, Oxford University Press, Oxford, UK (1991).

T. Brinck, J. S. Murray, and P. Politzer, J. Chem. Phys., **98**, 4305 (1993).

H. B. Callen, *Thermodynamics*, John Wiley & Sons, New York, USA (1960).

J. J. Gilman, "Shear-induced Metallization," Phil Mag., B, **67**, 207 (1993).

J J. Gilman, Mat. Res. Innovat., **1**, 71 (1997).

J. J. Gilman, Phil. Mag. A, **82**, #10, 1811 (2002).

J. J. Gilman, "Bond Modulus and Stability of Covalent Solids," Phil. Mag. Lett., **87**, 121 (2007).

J. C. Jamieson, Scuince, **139**, 845 (1963).

C. Kittel, *Introduction to Solid State Physics—7th Edition*, J. Wiley & Sons, New York, USA (1996).

R. G. Parr and W. Yang, *Density-Functional Theory of Atoms and Molecules*, Oxford University Press, New York, USA (1989).

R. G. Pearson, *Chemical Hardness*, Wiley-VCH, Weinheim, Germany (1997).

R. G. Pearson, Personal communication (2004).

J. A. Van Vechten, Phys. Rev. B, **7**, 1479 (1973).

W. Yang, R. G. Parr, and L. Uytterhoeven, Phys. Chem. Minerals, **15**, 191 (1987).

17 "Superhard" Materials

17.1 INTRODUCTION

Returning to a theme of Chapter 1, one of the factors that limits the advance of technology is the hardness of materials; alias strength. It limits the mechanical performance of many systems, such as: turbines, rocket engines, cutting tools, pressure vessels, energy storage devices, weapons, gyroscopes, aerospace structures, and the like. Thus there is strong continuing incentive to find harder materials, or effective combinations of hardness and other properties.

A benchmark for hardness is diamond, the hardest known substance. Its nominal hardness is 100 GPa (VHN = 10,000 kg/mm^2), but methods are known that may make it still harder. Based on this benchmark, materials with hardnesses between 20 and 40 GPa are said to be "very hard", while a material with hardness greater than 40 GPa is said to be "super-hard". The latter are very rare, and there is no true competitor for diamond. However, some property combinations make particular materials more useful than diamond in some applications. For example, cubic-BN is better for cutting iron-based alloys because it reacts chemically with Fe much less strongly than does the carbon of diamond. Therefore, its wear-rate is substantially less.

17.2 PRINCIPLES FOR HIGH HARDNESS

Hardness measures the resistance of a material to a permanent change of shape. That is, the resistance to shear deformation (not the resistance to a volume change). The precursor to a permanent shape change is a temporary elastic shape change, and a shear modulus determines this. Therefore, the first necessity for high hardness is a high shear modulus.

It is worth noting what determines elastic resistance to shear. Both shape changes and volume changes are determined by the behavior of the valence electrons in materials, but in quite different ways. Volume changes affect the average distances between the electrons, and between the valence electrons and their associated positive nuclei. Shear changes have little, or no, effect on these average distances because small shears do not affect volumes. However, shear causes a shift in the centroid of the electrons relative to the nucleus.

Chemistry and Physics of Mechanical Hardness, by John J. Gilman
Copyright © 2009 John Wiley & Sons, Inc.

This induces polarization and is resisted by a restoring force. The restoring force is essentially the same as the force that resists electric polarization and is determined by the polarizability (Gilman, 1997).

In terms of polarizability, α the shear elastic modulus, G is given approximately by:

$$G \approx (3/4\pi)\left(q^2/\alpha r\right) \tag{17.1}$$

where q = electron charge, and r = radius of an equivalent atom. Here α is not an atomic polarizability, but is the local polarizability in a crystal where it is a tensor quantity. As was shown in Chapter 9, the shear moduli of the alkali halides are inversely proportional to their polarizabilities. The latter are determined from their optical properties.

Note that in Equation 17.1, $\alpha \approx 4\pi r^3/3$ so $G \approx q^2/r^4$. But, the bulk modulus, B is also proportional to q^2/r^4, so G is proportional to B (approximately). Therefore, B can be used as a preliminary screen in a search for materials with high values of G.

High shear moduli alone do not equate with high hardness. A second require-ment is low dislocation mobility (high Chin-Gilman parameter). In other words, high resistance to plastic deformation. This property is associated with highly localized chemical bonds (covalent bonding) rather than non-local bonding (ionic and metallic bonding). A good example is the case of pure osmium metal.

Among the metals of the Periodic Table, osmium has the highest bulk modulus (412 GPa), and shear stiffness constants of $C_{44} = 270$ GPa and $C_{66} = 268$ GPa. (Pantea et al., 2008). The corresponding values for diamond are: B = 433 GPa and C_{44} (111) = 507 GPa. Although the bulk modulus of Os is about 95% that of diamond, the indentation hardness is only about 3% of diamond's hardness. In other words dislocations move readily in Os but not in diamond.

Dislocation motion can be impeded in many metals by introducing covalent bonds. A well-known case is that of titanium where TiB_2 is about 40 times harder than Ti alone. The B-atoms form covalent bonds. In the case of Os forming OsB_2 by adding B to the metal increases the hardness by about 10× (Cumberland, 2005QQ), and for Re the ratio is larger; about 15× (Chung et al., 2007).

Other methods for impeding dislocation motion are the introduction of grain boundaries, and/or twin boundaries. While these impediments may increase the hardness, they are also likely to decrease the tensile strength.

17.3 FRICTION AT HIGH LOADS

Friction forces between indenters and specimens are significant at all loads, but particularly for very hard specimens where the contact surfaces may "seize" if adsorbed gases get squeezed out of the contact zone. As discussed in Chapter 2, the effect of friction becomes increasingly important for small indentations (high hardness values). This causes the "indentation size effect". Friction is, of course, dependent on the friction coefficient which depends on both surface roughness and contact pressure (Mueser, 2008). For high values of these the coefficient can considerably exceed unity, and approach indefinitely large values.

Recent interest has developed in "nanocrystalline" aggregates which are apparently harder than their constituent crystals. However, the surfaces of such aggregates cannot be expected to be smooth; either geometrically or elastically. Therefore, being rough, they may have high friction coefficents, and the high values of hardness reported for them may be misleading. For diamond (Sumiya, et al., 2004). For cubic BN (Dubrovinskaia et al., 2007).

There is disagreement in the literature about the role of friction. Compare, for example, Cai (1993) with Ishikawa et al. (2000) This has arisen in various ways. In the case of metals, where the Chin-Gilman parameter is small, friction is not important for relatively large indents. However, as the C-G parameter becomes much larger for covalent crystals, and as the indent size decreases friction becomes more important. Also, environmental factors, such as humidity, affect friction coefficients. In the regime of superhardness with dry specimens and small indents friction becomes very important.

17.4 SUPERHARD MATERIALS

If superhardness is defined as $H > 40\,GPa$ ($4000\,kg/mm^2$) there are only a few cases:

Diamond	$H = 90–100\,GPa$	
BN	50	
B_6O	45	He et al., 2002
ReB_2	40	Chung et al., 2007
BC_2N	70	Solozhenko et al., 2001
$B_{13}C_2$	42	Domnich et al., 1998
B_3Si	54	Samsonov and Latysheva, 1955
B_5SiC_2	70	Portnoi et al., 1959

It is interesting that all of these crystals except diamond are boron compounds. Note also, that most of them consist exclusively of relatively small atoms. The exception is ReB_2. Since Re has a large number of valence electrons the general rule is followed that high hardness is associated with high VED (valence electron density).

Theoretically C_3N_4 would be superhard if it could be synthesized (Sung and Sung, 1996). However, its synthesis has not yet been achieved.

Reports of superhard composities (Veprek et al., 1998) must be viewed with skepticsm because of the effect of high friction on hardness measurements.

The same skepticism applies to nanocrystalline diamond that is reported to be harder than diamond single crystals (Sumiya and Irifune, 2007). Other cases in which rough surfaces may have skewed the measurements are TiN/SiN coatings (Kauffmann et al., 2005); and ($AlMgB_{14} + TiB_2$) mixtures (Cook et al., 2000).

Diamond itself can be hardened somewhat through plastic deformation. DeVries (1975) found that the wear resistance of diamond can be increased by compressing it under high confining pressure at 1200 °C. This hardening effect has been confirmed and extended with CVD diamonds (Yan et al., 2004). The latter authors measured hardness values as large as 160 GPa. Although the high values may have been influenced by the effects of friction (**17.3** above), there is no doubt that some hardening did result from their heat-"pressure" treatments (2000 °C—5–7 GPa—10 min).

REFERENCES

X. Cai, "Effect of Friction in Indentation Hardness Testing: A Finite Element Study", Jour. Mater. Sci. Lett., **12**, 301 (1993).

H-Y. Chung, M. B. Weinberger, J. Levine, A. Kavner, J-M. Yang, S. H. Tolbert, and R. B. Kaner, "Synthesis of Ultra-Incompressible Superhard Rhenium Diboride at Ambient Pressure", Science, **316**, 436 (2007).

B. A. Cook, J. L. Harringa, T. L. Lewis, and A. M. Russell. "A New Class of Ultra-hard Materials Based on $AlMgB_{14}$", Scripta Mater., **42**, 597 (2000).

R. W. Cumberland, M. B. Weinberger, J. J. Gilman, S. M. Clark, S. H. Tolbert, and R. B. Kaner, "Osmium Diboride, An Ultra-Incompressible, Hard Material", J. Am. Chem. Soc., **127**, 7264 (2005).

R. Devries, "Plastic Deformation and 'Work-hardening' of Diamond", Mat. Res. Bull., **10**, 1193 (1975).

V. Domnich, Y. Gogotsi, M. Trenary, and T. Tanaka, "Nanoindentation and Raman Spectroscopy Studies of Boron Carbide", Appl. Phys. Lett., **81**, 3783 (2002).

N. Dubrovinskaia, V. L. Solozhenko, N. Miyajima, V. Dmitriev, O. O. Kurakevych, and L. Dubrovinsky, "Superhard Nanocomposite of Dense Polymorphs of Boron Nitride: Noncarbon Material has Reached Diamond Hardness", Appl. Phys. Lett., **90**, 101912 (2007).

D. He, Y. Zhao, L. Daemen, J. Qian, and T. D. Shen, Appl. Phys. Lett., **81**, 643 (2002).

F. Kauffmann, B. Ji, G. Dehm, H. Gao, amd E. Arzt, "A Quantitative Study of the Hardness of a Superhard Nanocrystalline Titanium Nitride/Silicon Nitride Coat-ing", Scrip. Mater., **52**, 1269 (2005).

M. H. Mueser, "Rigorous Field-Theoretical Approach to the Contact Mechanics of Rough Elastic Solids", Phys. Rev. Lett., **100**, 055504 (2008).

K. I. Portnoi, G. V. Samsonov, and L. A. Solonnikova, "Reaction of Boron Carbide with Silicon", Dokl. Akad. Nauk SSSR, **125**, 823 (1959).

G. V. Samsonov, and V P. Latysheva, "Chemical Compounds of Boron with Silicon", Dokl. Akad. Nauk SSSR, **105**, 499 (1955).

V. L. Solozhenko, D. Andrault, G. Fiquet, M. Mezouar, and D. C. Rubie, "Synthesis of Superhard Cubic BC_2N", Appl. Phye. Lett., **78**, 1385 (2001).

H. Sumiya and T. Irifune, "Hardness and Deformation Microstructures of Nano-polycrystalline Diamonds Synthesized from Various Carbons under Hig Pressure and High Temperature", J. Mater. Res., **22** (8), 2345 (2007).

C-M. Sung and M. Sung, "Carbon Nitride and Other Speculative Superhard Materials", Mater. Chem & Phys., **43**, 1 (1996).

S. Veprek, P. Nesladek, A. Niederhofer, and F. Glatz, "Search for Superhsrd Materials: Nanocrystalline Composites with Hardness Exceeding 50 GPa", NanoStructured Mater., **10**, 679 (1998).

C. Yan, H. Mao, W. Li, J. Qian, Y. Zhao, and R. J. Hemley, "Ultrahard Diamond Single Crystals from Chemical Vapor Deposition", Phys. Stat. Sol. A, **201**, R25, (2004).

J. J. Gilman, "Chemical and Physical Hardness", Mater. Res. Innovat., **1**, 71 (1997).

M. Ishikawa, S. Okita, N. Minami, and K. Miura, "Load Dependence of Lateral Force and Energy Dissipation at NaF(001) Surface", Surf. Sci., **445**, 488 (2000).

C. Pantea, I. Stroe, H. Ledbetter, J. B. Betts, Y. Zhao, L. L. Daemen, H. Cynn, and A. Migliori, "Osmium's Debye Temperature", Jour. Phys. Chem. Sol., **69**, 211 (2008).

S. Sumiya, T. Irifune, A. Kurio, S. Sakamoto, and T. Inoue, "Microstructure Features of Polycrystalline Diamond Synthesized Directly from Graphite under Static High Pressure", Jour. Mater. Sci., **39**, 445 (2004).

INDEX

Chemistry and Physics of Mechanical Hardness, by John J. Gilman
Copyright © 2009 John Wiley & Sons, Inc.